Reliability of MEMS

Volume Editors
Osamu Tabata
Toshiyuki Tsuchiya

Related Titles

Hsu, T.

MEMS and Microsystems
**Design, Manufacture, and Nanoscale
Engineering, Second Edition**
Second Edition

2008
Print ISBN: 978-0-470-08301-7

Saile, V., Wallrabe, U., Tabata, O.,
Korvink, J.G. (eds.)

LIGA and its Applications

2009
Print ISBN: 978-3-527-31698-4; also available
in electronic formats

Hierold, C. (ed.)

Carbon Nanotube Devices
**Properties, Modeling, Integration and
Applications**

2008
Print ISBN: 978-3-527-31720-2; also available
in electronic formats

Bechtold, T., Schrag, G., Feng, L. (eds.)

System-level Modeling of
MEMS

2013
Print ISBN: 978-3-527-31903-9; also available
in electronic formats

Korvink, J.G., Smith, P.J., Shin, D.
(eds.)

Inkjet-based
Micromanufacturing

2012
Print ISBN: 978-3-527-31904-6; also available
in electronic formats

Baltes, H., Brand, O., Fedder, G.K.,
Hierold, C., Korvink, J.G., Tabata, O.
(eds.)

Enabling Technologies for
MEMS and Nanodevices
Advanced Micro and Nanosystems

2004
Print ISBN: 978-3-527-33498-8; also available
in electronic formats

Brand, O., Fedder, G.K. (eds.)

CMOS-MEMS

2013
Print ISBN: 978-3-527-33499-5; also available
in electronic formats

Kockmann, N. (ed.)

Micro Process Engineering
**Fundamentals, Devices, Fabrication,
and Applications**

2006
Print ISBN: 978-3-527-33500-8; also available
in electronic formats

Iannacci, J.

Practical Guide to RF-MEMS

2013
Print ISBN: 978-3-527-33564-0; also available
in electronic formats

Reliability of MEMS

Testing of Materials and Devices

Edited by
Osamu Tabata
Toshiyuki Tsuchiya

WILEY-VCH

The Volume Editors

Prof. Toshiyuki Tsuchiya
Kyoto University
Dept. of Micro Engineering
Yoshida Honmachi, Sakyo-ku
606-8501 Kyoto
Japan

Prof. Dr. Osamu Tabata
Kyoto University
Dept. of Micro Engineering
Yoshida Honmachi, Sakyo-ku
606-8501 Kyoto
Japan

Series Editors

Oliver Brand
School of Electrical and Computer Engineering
Georgia Institute of Technology
777 Atlantic Drive
Atlanta, GA 30332-0250
USA

Prof. Dr. Gary K. Fedder
ECE Department & Robotics Institute
Carnegie Mellon University
Pittsburgh, PA 15213-3890
USA

Prof. Dr. Jan G. Korvink
Institute for Microsystem Technology (IMTEK)
Albert-Ludwigs-Universität Freiburg
Georges-Köhler-Allee 103
79110 Freiburg
Germany

Prof. Dr. Osamu Tabata
Kyoto University
Dept. of Micro Engineering
Yoshida Honmachi Sakyo-ku
606-8501 Kyoto
Japan

■ First Edition 2007

All books published by Wiley-VCH are carefully produced. Nevertheless, authors, editors, and publisher do not warrant the information contained in these books, including this book, to be free of errors. Readers are advised to keep in mind that statements, data, illustrations, procedural details or other items may inadvertently be inaccurate.

Library of Congress Card No.:
applied for

British Library Cataloguing-in-Publication Data
A catalogue record for this book is available from the British Library.

Bibliographic information published by the Deutsche Nationalbibliothek
The Deutsche Nationalbibliothek lists this publication in the Deutsche Nationalbibliografie; detailed bibliographic data are available on the Internet at <http://dnb.d-nb.de>.

© 2013 Wiley-VCH Verlag GmbH & Co. KGaA, Boschstr. 12, 69469 Weinheim, Germany

Print ISBN: 978-3-527-33501-5
ePDF ISBN: 978-3-527-67503-6

Cover Design Grafik-Design Schulz, Fußgönheim

Typesetting Toppan Best-set Premedia Limited, Hong Kong

Printing and Binding Strauss GmbH, Mörlenbach

Printed in the Federal Republic of Germany
Printed on acid-free paper

Preface

Reliability is the ability of a system or component to perform its required functions under stated conditions for a specified period of time. For commercial products, reliability is one of the most important features. Since reliability often dominates the device designs, its maximum performance may be traded off in return for reliability. Reliability evaluation and control, however, also in turn contributes to the improvement of device performance. At the present day, industrial products are distributed all over the world and used in a broad range of environments, making reliability evaluation of the products more important than ever.

MEMS devices, which are micromechanical devices fabricated using semiconductor fabrication technologies, are core products for future integrated systems for miniaturization and advanced functionalities. Due to their compactness and portability, MEMS devices are being employed even in mobile applications. The reliability of MEMS devices are thought to be high because of the size effect with the miniaturization of their dimensions and the precision of micro-fabrication, which is confirmed by numerous commercially available devices, such as pressure sensors, accelerometers, inkjet printer heads, and projection displays. Although their high reliability has been established by the continuous development effort of many researchers and engineers, numerous reliability issues and their related phenomena still remain, and considerable effort is being spent to find a generalized theory on MEMS reliability both in devices and materials. In the near future, the knowledge on the MEMS reliability will be organized into a whole and every engineer will be able to design MEMS device of high performance and high reliability through this knowledge.

The mechanical reliability of micromaterials for MEMS structures is the principle part of the MEMS reliability, being the entirely new aspect which was not as important by far in classic microelectronic devices. The development of evaluation methods, experimental procedures and analysis of results measured has been forcefully driven forward in the past decade, resulting in much significant new knowledge.

In this volume of Advanced Micro and Nanosystems, entitled "MEMS Reliability," we aim to summarize and clarify latest cutting-edge knowledge on the mechanical reliability evaluation methods, their measurement results and use of measured data towards reliability and performance enhancement.

Reliability of MEMS.
Edited by O. Tabata, T. Tsuchiya
Copyright © 2013 WILEY-VCH Verlag GmbH & Co. KGaA, Weinheim
ISBN: 978-3-527-33501-5

The first part of the volume is devoted to mechanical property evaluation methods and their contribution to reliability assessment. Chapter one is of introductory nature, explaining to the reader the relationship between MEMS reliability and mechanical properties, and standardization of measurements, respectively. Chapters two, three and four cover the measurement methods, featuring nano-indentation, the bulge method and the uni-axial tensile test for determining fundamental mechanical properties of micromaterials. Reliability evaluation using device-like structures is described in chapter five, which is a useful method to evaluate the effects of fabrication methods and particular device design on the reliability properties.

The second part of the book gives a broad overview of MEMS devices which are highly reliable commercial products to demonstrate by example the importance of reliability assurance during product development. In chapters six, seven and eight the most successful MEMS-based mechanical sensors – pressure sensors, accelerometers, and angular rate sensors (vibrating gyros) – are described in detail. Particular emphasis is on the packaging and assembling, demonstrating the importance of these steps for device reliability. In chapters nine and ten, we turn to optical MEMS devices such as variable optical attenuators for fiber optical communication and two-dimensional optical scanners, both operated in torsional deformations. These optical devices operated in high frequency, mostly in resonance, are requested to have long lifetime in their operations because of their applications, such as telecommunication system, safety system, and display system.

Another aim of this volume is to introduce the reader to MEMS research and development in Japan. The Japanese industry possesses considerable MEMS device knowhow which has not yet been globally presented in appropriate detail. It is my hope that MEMS researchers and engineers in the world may profit and build upon these efforts and contributions.

Finally, I would like to thank all the contributors for their kind collaboration. They spent considerable time to write their chapters despite numerous other commitments and full schedules. In turn, I hope the readers may enjoy the book and gain further insights and understanding of MEMS reliability through this volume.

Toshiyuki Tsuchiya
Kyoto, June 2007

Foreword (by Series Editors)

We are proud to present the sixth volumes of *Advanced Micro & Nanosystems* (*AMN*), entitled *"Reliability of MEMS"*.

In the past two decades, after the concept of MEMS was propounded, various kinds of micro-devices have been developed and commercialized. As symbolized by the term MEMS – the acronym of "Micro Electro Mechanical Systems" – the behavior of the mechanical structure plays an important role in the operation of these devices. A complete understanding of the relevant mechanical properties of thin films and the resultant mechanical behavior of microstructures is an utmost priority in progressing on the broader range of MEMS applications. This need has motivated research on methodologies to measure and evaluate the mechanical behavior and mechanical properties. Through these untiring efforts, many useful methodologies have been developed and relevant data has been accumulated. Reliability related material properties, such as strength and fatigue, have much more importance than ever before. The need for increased reliability studies is not only for the many MEMS products that are already commercialized but also for emerging MEMS applications operating in harsh environments such as high temperature, high pressure and high radiation. As one indication of this new research priority for reliability, in recent presentations at conferences in the MEMS field, reporting of device performance on long-term stability of behavior such as sensitivity, resonant frequency and noise figure is now much more prevalent than in the past. As represented by these facts, reliability of MEMS finally has become a performance property that should be assured through careful theoretical and experimental investigations.

However, there is still a long way to go to attain this goal. New findings, such as refined fatigue properties of single crystalline silicon, impose on the MEMS engineers to revise device design criteria. Though a large amount of data has been obtained from present devoted research efforts, comprehensive theory for describing the fatigue properties of MEMS devices and materials has not been established. The accumulated data and acquired knowledge should be shared by all the researches and engineers in MEMS to reach this goal.

In this volume, the mechanical reliability of MEMS, both from the material research and device development perspectives, is described to elucidate our knowledge about the reliability. The first part of this volume is devoted to material

Reliability of MEMS.
Edited by O. Tabata, T. Tsuchiya
Copyright © 2013 WILEY-VCH Verlag GmbH & Co. KGaA, Weinheim
ISBN: 978-3-527-33501-5

research on evaluation methods of mechanical properties and on the measured properties for MEMS devices. The second part describes recently developed MEMS devices and their reliability related properties. Through reading this volume, we hope all of you gain much more interest in the reliability of MEMS and share and extend the knowledge and help in the development of highly reliable MEMS devices. To accomplish such a goal, we are very glad to have the support of Prof. Dr. Toshiyuki Tsuchiya from Kyoto University, Japan, who is the editor of this volume.

Oliver Brand, Gary K. Fedder, Christofer Hierold, Jan G. Korvink, and Osamu Tabata
Series Editors
May 2007
Atlanta, Pittsburgh, Zurich, Freiburg and Kyoto

Contents

Reliability of MEMS.
Edited by O. Tabata, T. Tsuchiya
Copyright © 2013 WILEY-VCH Verlag GmbH & Co. KGaA, Weinheim
ISBN: 978-3-527-33501-5

List of Contributors

Changho Chong
Santec corporation
Applied Optics R&D group
5823 Nenjyozaka, Ohkusa,
Komaki, Aichi, 485-0802
Japan

Joao Gaspar
University of Freiburg
Georges-Koehler-Allee 103
79110 Freiburg
Germany

Keiji Isamoto
Santec corporation
Optical component design group
5823 Nenjyozaka, Ohkusa,
Komaki, Aichi, 485-0802
Japan

Harold Kahn
Department of Materials Science
    and Engineering
Case Western Reserve University
10900 Euclid Avenue
Cleveland, OH 44106-7204
USA

Kenji Komaki
Sumitomo Precision Products Co., Ltd.
1-10 Fuso-cho
Amagasaki
Japan

Takahiro Namazu
University of Hyogo
Department of Mechanical and
    Systems Engineering
Graduate School of Engineering
2167 Shosha
Himeji
Hyogo 671-2201
Japan

Oliver Paul
University of Freiburg
Georges-Koehler-Allee 103
79110 Freiburg
Germany

Mototsugu Sakai
Toyohashi University of Technology
Department of Materials Science
1-1 Hibarigaoka
Tempaku-cho
Toyohashi 441-8580
Japan

Reliability of MEMS.
Edited by O. Tabata, T. Tsuchiya
Copyright © 2013 WILEY-VCH Verlag GmbH & Co. KGaA, Weinheim
ISBN: 978-3-527-33501-5

Sho Sasaki
OMRON Corporation
Corporate Research and
Development Headquarters
9-1 Kizugawadai
Kizugawa-City
Kyoto 619-0283
Japan

Fumihiko Sato
OMRON Corporation
Corporate Research and
Development Headquarters
9-1 Kizugawadai
Kizugawa-City
Kyoto 619-0283
Japan

Osamu Tabata
Kyoto University
Department of Micro Engineering
Graduate School of Engineering
Yoshida-Honmachi
Sakyo-ku
Kyoto 606-8501
Japan

Osamu Torayashiki
Sumitomo Precision Products
   Co., Ltd.
1-10 Fuso-cho
Amagasaki
Japan

Hiroshi Toshiyoshi
University of Tokyo
Institute of Industrial Science
4-6-1 Komaba
Meguro-ku
Tokyo 153-8505
Japan

Toshiyuki Tsuchiya
Kyoto University
Department of Micro Engineering
Graduate School of Enginereing
Yoshida-Honmachi
Sakyo-ku
Kyoto 606-8501
Japan

Yuzuru Ueda
Nippon Signal Co., Ltd.
1836-1 Oaza Ezura
Kuki-shi
Saitama 346-8524
Japan

Hideaki Watanabe
OMRON Corporation
Corporate Research and Development
   Headquarters
9-1 Kizugawadai
Kizugawa-City
Kyoto 619-0283
Japan

Akira Yamazaki
Nippon Signal Co., Ltd.
13F Shin-Marunouchi Building
1-5-1, Marunouchi
Chiyoda-ku
Tokyo 100-6513
Japan

Noriyuki Yasuike
Matsushita Electric Works, Ltd.
1048 Kadoma
Osaka
Japan

Overview: Introduction to MEMS Reliability

Toshiyuki Tsuchiya[1] and Osamu Tabata[2]

[1]*Associate Professor, Department of Micro Engineering, Kyoto University, JAPAN*
[2]*Professor, Department of Micro Engineering, Kyoto University, JAPAN*

As the overview of the volume, "MEMS reliability", we describe the current under-standings on mechanical reliabilities of microelectromechanical system (MEMS) to contribute the expansion of the device applications. "MEMS reliability" has wide meanings in exact sense, because MEMS is complex systems containing wide range of physical and chemical theories. Since the fabrication technologies in MEMS are mostly based on that in semiconductor devices, the electrical properties evaluation can be utilized, such as thermal cycle test and accelerated life time test in high temperature. However, in MEMS we should consider the mechanical properties. In their reliability properties, various mechanical tests need to be oper-ated, such as shock survival and long term endurance in static and dynamic load-ings. In addition, MEMS consists mainly of silicon that is a brittle material and has not been considered as a mechanical structural material. Engineers think MEMS requires subtle treatment in their operations. In this overview, we focused on the mechanical reliability in MEMS, especially in devices consisted of silicon structures.

Mechanical reliability in MEMS

Microelectromechanical Systems (MEMS) means an integrated device fabricated using (silicon) micromachining, which contains mechanical structures, transduc-ers, and controlling and detecting circuit on single or multiple chips. The following definitions were provided by a research organization in the United States,

> "MEMS means batch-fabricated miniature devices that
> convert physical parameters to or from electrical signals
> and that depend on mechanical structures or parameters in
> important ways for their operation." [1]

Reliability of MEMS.
Edited by O. Tabata, T. Tsuchiya
Copyright © 2013 WILEY-VCH Verlag GmbH & Co. KGaA, Weinheim
ISBN: 978-3-527-33501-5

As clearly pointed in this definition, the mechanical properties of micro-materials in MEMS are strongly associated with the devices' performances as well as their mechanical reliability. The strong relationship between the performances and reliability exists and the reliability problem often limits the performances. The device should withstand to the shock, at same time it detects tiny inertial force caused by input acceleration.

The mechanical shock causes two severe reliability problems in structure of MEMS, which are stiction and fracture. Stiction means that two structures bonded together and never apart in controlled or unintentional contact between them. The stiction is critical in smaller structures because it is caused by relatively large interfacial force compared to the restoring force. The control of these two forces is crucial. Since the reduction of interfacial force is done by mainly chemical processing [2], we will not discuss in this chapter. For stiction, the controls of the contact area and surface conditions are important. In order to avoid the large deflection by the shock acceleration, device structures often have stoppers to limit the movement of the movable structure. The increase of restoring force can be done by the increase of the spring constant of the structure. However, it is not practical because the sensitivity decreases by the higher stiffness.

Fracture may occur when large force is applied to suspending structures and stress in structures exceeds its durable limit. Since structural components in MEMS are often made of brittle materials, risk of fractures both in static and dynamic loading was pointed out and investigated. The evaluation of mechanical reliability of micro-materials was proclaimed. This is one of the major motivations that research on the mechanical properties measurement has become active. The measurement of the elastic constants used for the performance design is another one. Through the previous research on mechanical properties measurement, a lot of issues for mechanical reliability evaluations are pointed out because of the size and features of the MEMS. Issues in mechanical reliability evaluations are described below.

- **Issues on mechanical reliability evaluation in MEMS**

Though a lot of mechanical reliability evaluations have been performed, it is difficult to discuss generalized theory about the reliability properties of MEMS because all of the evaluations are done by different methods and conditions. The differences in the methods are mainly caused by the lack of existence of standard test methods and their procedures. It is no doubt that there are difficulties for the establishment the standard test methods because of the variation of device properties. The differences in test methods are discussed later.

The decision of the test conditions in the reliability evaluation has many difficulties. For the dynamic fracture (fatigue) test, the various experimental conditions are in question, such as the stress and strain amplitude, stress ratio, test frequency and test cycles. The structures in most devices experience both tensile and compressive stress during their operations, which means the stress ratio is -1. However, it is difficult to apply uniform compressive stress in tiny specimen for fracture test by uni-axial tensile stress. The stress and strain control during the cyclic loadings is difficult due to the precise measurement of the small force and displacement or elongation.

The test frequency and cycles in fatigue test are the critical issues. In MEMS reliability evaluation, ultra high cycles of loading are requested because of their dimensions. We prefer to use vibrating tests to reduce time and cost of the reliability evaluations, but there is no theory of accelerating factor in MEMS structures. The lack of standard test procedure causes the difference in test conditions.

Analysis of the measured results and prediction of the lifetimes are still under investigation even for silicon structures, which is strongly related to the insufficient knowledge about the material properties of micro-materials. There is neither established numerical model nor lifetime estimation method. The lifetime of semiconductor devices are often evaluated using the accelerated lifetime test (ALT). In ALTs, accelerating parameters, such as the temperature, humidity, stress, and frequency in test condition, are needed to reduce the evaluation time and cost. However, the lack of fatigue mechanism in MEMS causes reliable ALT difficult.

Reliability evaluation methods

Various reliability evaluation methods have been employed in MEMS and a lot of research papers have been reported. However, the details of each evaluation were extremely different because the object or motivation of these evaluations directed to each research's interest, for example, products development, material science, and evaluation method development. In this section, to understand the differences in the reliability evaluation methods, the reliability test structure for MEMS are categorized into three levels, called as specimen-level, device-level, and product-level. Figure 1 illustrates the schematic description of these three levels. Through this categorization, we would like to point out what is the limit of the existing evaluation methods and future prospects for development of reliability evaluation methods and theoretical explanation of reliability phenomenon in MEMS.

• Specimen-level evaluation

In the first level of the reliability test, the test pieces has a simple beam like structures that looks like miniaturized dog bone specimen for standard tensile test or cantilever beam for bending test. Figure 2 shows specimens for uni-axial and bending test. The stress applied on the test pieces is simple, such as uni-axial or bi-axial tensile and bending stress. This type of test pieces is useful for stress and stains extraction and is used for material research. Examples of the evaluation in this level are described in Chapters three and four.

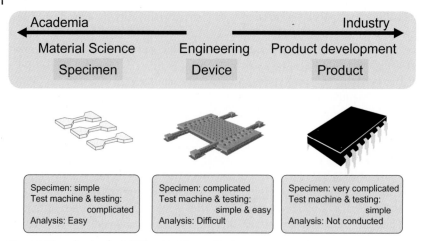

Figure 1 Three levels of reliability evaluation methods.

Figure 2 Test pieces for "specimen" level reliability evaluation methods. Left: single crystal silicon specimens for uni-axial tensile testing. Left end of the test pieces are fixed to substrate and right end is chucked by adhesive glues or electrostatic force. Right: polysilicon specimens for cantilever beam bending test.

Uni-axial tensile tests were performed as a miniaturization of the bulk tensile test. The fracture force is relatively large compared with this dimensions. For example, 2-µm-wide and 2-µm-thick single crystal silicon fractures with tensile load of about 10 mN, which is too large to chuck and load using device structures and actuators. Many experiments adopted a special and dedicated specimen chucking or loading methods, which is described in detail in Chapter four. As the result, the skill for test is often required. In addition, the fabrication process of the specimens is often special and unusual, which prevents us to use the results directly to products.

- **Device-level evaluation**

The second type of reliability test pieces is an on-chip structure often integrated with loading and detection transducers. The structure is similar to or same as the commercial device, but often simplified for easy extraction of stress and strain, which is discussed in detail in Chapter five. The device is often operated in resonance, because only small actuation force can be generated for initiating fracture. In addition, notch or crack introduction is crucial for generating stress concentration. An example of the level is shown in Figure 3.

Device-level evaluation is simple and easy. The useful data for designing the product using similar device structure to the test piece will be obtained. However, the analysis of the results in material science may be often difficult because of the complicated structure and stress concentration. For the future, the device-level evaluation is necessary for interpretation of the products reliability phenomenon using material reliability properties gathered by the specimen-level evaluation.

- **Product-level evaluation**

In the third type of reliability test pieces, tests are carried out on the device to be commercially available, which we call it "product-level". Test conditions are often assigned to emulate the actual environment. Analysis of the results for reliability assessment is easy. This type of reliability test is mainly for industries. Direct reli-

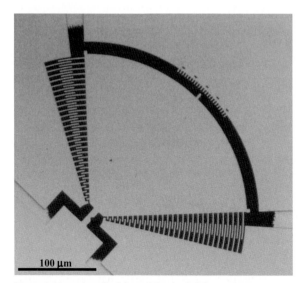

Figure 3 Test piece for "device" level reliability evaluation method. Fan shaped mass made of single crystal silicon suspended by notched beam. Two sets of comb drive actuators are used for oscillation and detection of mass vibration.

ability information can be gathered, which is useful for the engineer who develops the device, but is often useless for others and even for the engineer when he changed the device design. In addition, the reliability test results seldom publish or present to public, even if the device is commercially available. The latter part of this book represents the reliability test results that are valuable.

Each level of reliability test has been solely carried out. No relationships between the levels have been established, because too many parameters contribute for mechanical properties. The device level reliability test with consideration on the test parameters of both specimen-level and product-level tests is important for merging various kinds of reliability test results. The fracture and fatigue mechanism investigation to each level test piece is required.

Silicon

Silicon is a basic, common, and most widely used material in MEMS. They have been already employed in various control and measurement systems, even in the systems that require high reliability such as automotive application. However, it is still unknown whether silicon is mechanically reliable or not. In early years, the reliability of a bulk micromachined silicon accelerometer was assured using a burn-in test, in which device is initially loaded at a stress much higher than that in its normal operation [3]. If the device survives in this test, it can conclude that the devices will not fail forever, because defects in silicon do not move at room temperature and thus shows no plastic deformation. In recent years, many researches on the mechanical reliability of silicon in mainly surface-micromachined device structures showed that it fails during constant stress applications [4, 5]. In addition, industrial people are saying that no one knows whether the structure will fail or not on the next day after ten years successful operation.

Various experimental works on the fatigue properties measurement and phenomenon analysis of silicon showed its distinctive properties, such as the large deviation, environmental effect, and resonant frequency changes.

• Large deviation

Large deviations in strength and fatigue life were observed, which is mainly caused by the fabrication process. The reported strength by static and dynamic loadings often shows large difference between the test methods and low repeatability in the same test methods. In addition, the deviation in the strength is large. Silicon has been widely considered as a delicate material. However, if we can control the fabrication process the repeatable results can be obtained. Chapter one describes the cross-comparison of tensile testing methods, which shows that the mechanical properties measured by the different testing methods were same for the simultaneously fabricated specimen on the same wafer. We also observed the fatigue properties of single crystal silicon resonator fabricated using reduced projection lithography which has small deviation in fatigue strength [6]. This result also indicates that the precise and accurate fabrication allows us to make highly reliable devices.

- **Environmental effect**

The effect on humidity was demonstrated in both the cantilever beam[4] and doubly supported beam resonators [7]. Considering that the fracture is initiated at the surface defects, the fatigue fracture is caused by the growth or widen of defects during the long-term stress application [8]. It is well agreed that the water vapor affects or initiates the defect growth. Native oxide layers on silicon surface are thought to play important role on the humidity effect on fatigue properties. Temperature is another important factor on fatigue properties. Though fatigue test in higher temperature has not been reported, the static strength in high temperature was reported in many papers. Ductile deformation was observed at relatively low temperature in micrometer thick silicon thin film, where bulk silicon shows ductility at least 600°C . The minimum temperature in which silicon shows ductility decreased with decreasing the specimen dimensions. In the fracture toughness test, significant increase of fracture toughness at 70°C was observed in micronsized single crystal silicon films [9].

- **Resonant frequency changes**

During the fatigue tests of MEMS devices using resonant vibration, the frequency changes were observed. The reported maximum frequency changes were 0–0.06% during 10^5–10^{11} cycles vibration in the polysilicon lateral rotational resonator, whose resonant frequency was about 42 kHz [8] and 1% during 10^7 cycles vibration in the single crystal silicon vertical resonator, whose resonant frequency was about 11 kHz [4]. The frequency changes had no relationship to the fracture occurrence, whereas their mechanisms were related to the fatigue failures.

These results suggested that the fracture of silicon is initiated from the surfaces and the native oxide layer and its growth plays an important role of reliabilities. Some theories have been proposed. In the "reaction layer fatigue" mechanism, native oxide layer is grown on the silicon surface and thickens at the stress concentrated position. In this oxide layer, small cracks are generated and further growth of oxide layer occurs where oxygen is supplied through this crack. By repeating this process, one of the cracks becomes the fracture origin to reach the critical length [10, 11]. The native oxide layer is so thin that no fatigue fracture is observed in bulk scale specimens. Lavan, et al. investigated the fatigue properties against stress amplitude and stress ratio as he describes in Chapter five. The stress amplitude is directly related to the fatigue failure. However the combination of static stress and small amplitude of cyclic stress caused strengthening of the structure.

There has been no consensus and stable theory in fatigue fracture of silicon. We need much more test results about silicon fatigue. Each chapter of this volume picks up mainly silicon as the object of reliability assessment. These works will contribute to establish the general theory of reliability of silicon.

Current status of knowledge about mechanical reliability evaluation in MEMS structures is summarized in this overview. A lot of evaluation methods have been developed and measurement results were reported, but a little knowledge has been revealed. As for the reliability, there is no consensus in the mechanism of "fatigue failure" in silicon. There are many things to do, such as the method development, standardization of test procedure, and physical explanations in mechanisms of reliability. Such works are presented in each chapter in this volume. We hope this overview helps you to read this volume.

References

1 http://www.wtec.org/loyola/mems/

2 R. Maboudian, C. Carraro, "Surface Chemistry and Tribology of MEMS", *Annual Review of Physical Chemistry* 54, (2004) 35–54.

3 M. Mutoh, M. Iyoda, K. Fujita, C. Mizuno, M. Kondo, M. Imai, Development of Integrated Semiconductor-Type Acceleration. *Proc. IEEE Workshop on Electronic Applications in Transportation*, 1990 pp.35–38.

4 J. A. Connally, S. B. Brown, Slow crack growth in single-crystal silicon. *Science* 256 (1992) 1537–1539.

5 S. Brown, W. V. Arsdell, C. L. Muhlstein, Materials Reliability in MEMDevices. *Proc. International Conference on Solid-State Sensors and Actuators* 1997 pp. 591–593.

6 T. Ikehara, T. Tsuchiya, High-cycle fatigue of micromachined single crystal silicon measured using a parallel fatigue test system. *IEICE Electron. Express* 4 (2007) 288–293.

7 T. Tsuchiya, A. Inoue, J. Sakata, M. Hashimoto, A. Yokoyama, M. Sugimoto, Fatigue Test of Single Crystal Silicon Resonator. *Tech. Digest of the 16th Sensor Symposium*, Kawasaki Japan, (1998) 277–280.

8 C. L. Muhlstein, S. B. Brown, R. O. Ritchie, High-cycle fatigue of single-crystal silicon thin films. *J. Microelectromechanical Systems* 10 (2001) 593–600.

9 S. Nakao, T. Ando, M. Shikida, K. Sato, Mechanical properties of a micron-sized SCS film in a high-temperature environment. *J. Micromechanics and Microengineering* 16 (2006) 715–720.

10 C. L. Muhlstein, S. B. Brown, R. O. Ritchie, High-cycle fatigue and durability of polycrystalline silicon thin films in ambient air. *Sensors and Actuators A* 94 (2001) 177–188.

11 O. N. Pierron, C. L. Muhlstein, The extended range of reaction-layer fatigue susceptibility of polycrystalline silicon thin films. *Int. J. Fracture* 135 (2005) 1–18.

1

Evaluation of Mechanical Properties of MEMS Materials and Their Standardization

Toshiyuki Tsuchiya, Department of Micro Engineering, Kyoto University, Japan

Abstract

The importance of the mechanical properties evaluation on designing and evaluating MEMS and the development of standard on MEMS are described in this chapter. First, in order to confirm their importance, the effect of mechanical properties on the performance of MEMS is pointed out. Second, to reveal the accuracy and repeatability of the existing evaluation methods, a work for cross comparisons is described. Then, the current workings on the international standard development on thin film mechanical properaties to improve the reliability, repeatability, and accuracy in the mechanical properties evaluation are introduced.

Keywords

thin films; mechanical properties; MEMS; tensile testing; standardization

Reliability of MEMS.
Edited by O. Tabata, T. Tsuchiya
Copyright © 2013 WILEY-VCH Verlag GmbH & Co. KGaA, Weinheim
ISBN: 978-3-527-33501-5

1.1
Introduction

Evaluations of the mechanical properties of micro- and nano-materials, especially thin films, which form mechanical structures of microelectromechanical system (MEMS) devices, are significant irrespective of the commercialization of applied devices for MEMS. The properties of thin films have been evaluated to satisfy demands in semiconductor device research, but they were mainly on the electrical properties. Studies on evaluations of mechanical properties have been limited, mainly to internal stresses. When the mechanical properties were needed, the bulk properties were often adopted, which was sufficient for their demands. However, when thin films started to be used for various mechanical structures, the mechanical and electromechanical properties play important roles in the operation of MEMS devices. Therefore, the mechanical properties of thin films need to be measured, and accurate properties similar to the electrical properties in semiconductor devices are required.

The mechanical properties of thin films should be measured on the same scale as micro- and nano-devices, since they are different from those of bulk materials. Reasons of the differences are follows;

- *Size effects:* The ratio of the surface area to the volume increases with decrease in the dimensions of a device structure. The surface effect might be more effective in MEMS devices. For example, the fracture of silicon, a brittle material, was initiated from the surface defects that are mainly produced during the fabrication process and the surface roughness dominates the strength. The size effect would be more sensitive at the microscale.
- *Thin-film materials:* Thin-film materials often have different compositions, phase and microstructure from the bulk materials, even if they are called by the same material names. The formation processes, such as deposition,

thermal treatment, implantation and oxidation, are inherent methods for thin-film materials. For example, bulk "silicon nitride" is a polycrystalline material and often contains impurities for improving properties, but silicon nitride thin films are deposited by chemical vapor deposition and are amorphous and seldom doped by impurities.

- *Processing:* Mechanical processing, which is the most commonly used processing method for bulk structure, is rarely used because the processing speed is too fast for the microscale. Instead, photolithography and etching are widely used. The surface finishing of the processed structure is completely different between the bulk and thin film.

These are the reasons for the necessity for the direct measurement of thin-film materials. In addition, it reveals the effects of their formation, processing and dimensions on their mechanical properties. The dimensions of the structures in MEMS devices have wide ranges, from sub-micrometers to millimeters. Evaluations of the mechanical properties of thin films cover a very wide range of measurement scale. Many measurement methods have been developed and various values have been measured using these methods.

However, studies on both the development of measurement methods and the evaluation of thin-film materials showed that there were inaccuracies in the measured results obtained by each method. The variations in the measured properties were large but the source of the variation was not established since there were too many differences among the properties measured by the different methods. The accuracy of the measurement methods, which is the basis of the evaluation, has not been verified because there are no standards for the mechanical properties of thin films. Recently, the development of international standards for measurements of mechanical properties was initiated in order to obtain more accurate properties and reliable measurements.

In this chapter, the importance of the mechanical properties for MEMS devices is defined to confirm the necessity for the evaluation of method developments and their standardization. The effects of each mechanical property on the design and evaluation of the devices are pointed out. Then, cross-comparisons of the evaluation methods for mechanical properties are described to indicate the critical points for more accurate measurements at the thin-film scale. Finally, current progress in the development of international standards on thin-film mechanical properties to improve the reliability, repeatability and accuracy of measurements of mechanical properties is discussed.

1.2
Thin-film Mechanical Properties and MEMS

The evaluation of the mechanical properties of thin films is indispensable for designing MEMS devices, since the properties play the following roles:

- *Device performance:* In MEMS devices, the mechanical properties are closely connected to the device performance. Accurate values of the mechanical properties are needed for obtaining the best performance.
- *Reliability:* MEMS devices are intended to be used in harsh environments because of their small size. Reliability is one of the most important properties.

In addition, the establishment of a properties database is required in order to accumulate knowledge about design information. Recently, the rapid prototyping of MEMS devices by incorporating MEMS foundry services and CAD/CAE software dedicated to MEMS devices has attracted much interest. A suitable database of thin-film mechanical properties should be compiled in order to ensure the most appropriate designs. This section provides descriptions of the effect of each mechanical property on the properties of MEMS devices to emphasize the importance of their evaluation.

1.2.1
Elastic Properties

Elastic properties, such as Young's modulus, Poisson's ratio and shear modulus, are directly related to the device performance. The stiffness of a device structure is proportional to the Young's modulus or shear modulus and the resonant frequency is proportional to the square root. However, as discussed in the next section, the stiffness of the thin-film structure depends additionally on the internal stress and the internal stress changes by an order of the magnitude, and these effects of the elastic properties on the device performance should be considered as a maximum effect. The acceptable errors in the elastic properties will be a few percent for cantilever beam structures and folded beam structures in which the internal stress has no effect on the stiffness of the structure. Larger deviations will be acceptable for structures whose internal stress dominates their stiffness. The stiffness of membrane structures for pressure sensors and diaphragm pumps is affected by the Poisson's ratio. The pressure P and center deflection w_0 of a circular membrane, as shown in Figure 1.1(a), are expressed by

$$P = \frac{4\sigma_0 h}{r^2} w_0 + \frac{8Eh}{3(1-v)r^4} w_0^3 \tag{1}$$

where h, r, E, σ_0 and v are the thickness, radius, Young's modulus, internal stress and Poisson's ratio of the membrane, respectively. The range of the Poisson's ratio of materials is not wide and the effect is not large, as shown in Eq. (1). A rough estimation of the Poisson's ratio by using the bulk properties is often acceptable.

These arguments will lead to the conclusion that the temperature coefficients of the elastic properties are negligible for most sensor devices. For a specific application, such as oscillators and filters which use MEMS structure as resonators, the deviation and the temperature coefficient should be more precisely measured and controlled. They require stability of the resonant frequency of the order of ppm.

1.2.2
Internal Stress

The internal stress, the strain generated in thin films on thick substrates, is not an elastic property in the strict sense. If the stress is present along the longitudinal direction for a doubly supported beam structure and the in-plane direction for a fixed-edge membrane structure, the stiffness along the out-of-plane direction of the structure has terms of the internal stress. Since the internal stress has an effect similar to large displacement analysis, the effect of internal stress on the stiffness and resonant frequency should be considered as closely as that of the Young's modulus.

The doubly supported beams shown in the Figure 1.1(b), and also the membranes shown in Figure 1.1(a), are loaded by the internal stress, hence the stiffness will change. The lateral stiffness of the doubly supported beam structure shown in Figure 1.1(b) is described by

$$F = \frac{4Ebh^3}{l^3}\left[1 + \frac{\pi^2\sigma_0}{8E}\left(\frac{l}{h}\right)^2 + \frac{\pi^4}{128}\left(\frac{w_0}{h}\right)^2\right]w_0 \tag{2}$$

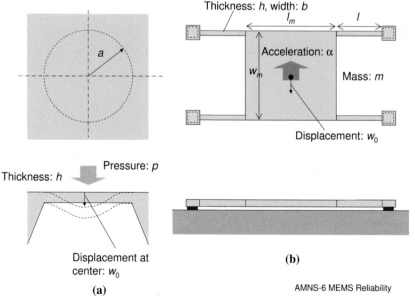

Figure 1.1 Typical structures of MEMS devices. (a) Thin diaphragm for pressure sensors; (b) doubly supported mass-beam structure for resonator and accelerometer.

where l and w are the length and width, respectively, of the beams. If the center mass consists of thin films and the stress in the mass is released, the additional stress is applied on the supporting beam. In this case, the modified stress is described by

$$\sigma_c = \sigma_0 \left(1 + \frac{l_m}{2l}\right) \tag{3}$$

where the l_m is the mass length.

The range of the internal stress values is wide; in the case of polysilicon, the stress range is from −500 to 700 MPa depending on the deposition methods and conditions and heat treatments. The negative (compressive) stress causes a decrease in the stiffness. Zero stiffness leads to the buckling of the structure. Stress control and accurate measurement are more important factors.

The origin of internal stress is classified into the intrinsic stress and the thermal stress. Chemical reactions, ion bombardment, absorption and adsorption cause the intrinsic stress, which can be controlled by the deposition conditions. However, control of the repeatability of the process conditions and resulting internal stress is very difficult. The thermal stress is caused by the mismatch of the coefficients of thermal expansion of the thin film and of its substrate. The thermal stress often becomes the origin of the temperature properties of the structures; the release or control of the thermal stress should be considered in designing device structures.

The internal stress may cause the destruction of the structure. High compressive stress causes buckling as discussed above and high tensile stress causes the fracture of structures. In both cases, film peeling is possible with large stresses.

The internal stress considered above is assumed to be uniform along the thickness direction. Actually, the stress is often distributed along the thickness direction, which causes out-of-plane deflection of cantilever structures.

1.2.3
Strength

The strength of the thin-film materials need to be evaluated and controlled to assure and improve the reliability of MEMS devices. The strength is the main parameter for the deposition process, etching, microstructures and shape uniformity. These parameters should be considered in order to evaluate the reliability. When engineers apply the measured strength values to their own devices, they should consider not only the test methods but also the fabrication methods of the specimens. For example, on designing the strength of a membrane structure which is to be used for a pressure sensor and has no etched surfaces, the tensile or bending strength of cantilever beam specimens should never be used because the beam structure has etched surfaces, dominating the fracture properties. In addition, the loading direction should be considered when evaluating beam structures. The lateral and vertical strengths may be different even if the same specimen is tested.

MEMS devices are expected to be used in mobile and portable applications, where the system and device structures are expected to have high durability against shocks. The requirement for shock durability often causes the difficult device design because the stress generated by the shock is larger than the stress applied in their normal operation. For example, accelerometers for automobile and mobile applications are designed to have measurement ranges of few to few tens of G (gravity). However, the shock applied with a drop from a height of 1.5 m on to a concrete surface is said to be equivalent to 3000–10 000 G. If the shock is applied directly, the device structure will have a stress of at least 100 times larger than the stress due to the designated input of the sensor. Therefore, the device has a stopper as a shock reduction structure. When there is no stopper structure because of the fabrication capability, the device sensitivity is limited to reducing the stress during shock.

In the case of vibrating gyroscopes that measure the Coriolis force to sense the angular rate, the shock can be reduced by adding a damper in the packaging structures. Accelerometers do not have such a damper because it causes a reduction in response time.

1.2.4
Fatigue

Fatigue is observed as a change in elastic constants, plastic deformation and strength decrease through the application of a cyclic or constant stress for a long time. Plastic deformation and changes in elastic constants cause sensitivity changes and offset drift in devices and fatigue fracture causes sudden failure of the device functions. These should be avoided in order to realize highly reliable devices.

Silicon, the most widely used structural material, shows no plastic deformation at room temperature. In addition, silicon was thought to show no fatigue fracture, which means that it suffers no decrease in strength on long term application of stress. Therefore, previously some engineers did not consider the fatigue of silicon MEMS devices. However, various experiments have shown fatigue fracture and decrease in strength of more than few tens of percent of the initial strength. Now all MEMS engineers consider the reliability of silicon structures to increase the device reliability.

Metal films, such as aluminum and gold, which are used in micromirror devices, show plastic deformation and metallic structures may show large drift and changes in performances. For example, digital micro mirror devices (DMDs) [1] are operated by on–off state, which is acceptable for change in material properties.

As for the size effect of MEMS structures, the surface effect will contribute greatly to the fatigue properties. The effect of the environment, such as temperature and humidity, should be evaluated. The resonant frequencies of the MEMS structures are higher than those of macroscale devices. In order to assure the long-term reliability of such devices, we should evaluate the reliability against a large number of cyclic stress applications. If we assume that the resonant frequency of the device is about 10 kHz, 1 year of continuous operation equals 10^{11-12}

cycle loadings. Therefore, proper accelerated life test method and life prediction method are required by analyzing the mechanism of the fatigue behavior.

1.3
Issues on Mechanical Properties Evaluations

In the design of MEMS devices and confirmation of their reliability, evaluations of the mechanical properties of micro- and nano-materials are crucial. However, there are some issues regarding accurate measurements, which are related mainly to the accuracy of the measured values.

1.3.1
Issues Related to Specimens

Deviations of specimen dimensions are one of the most important and basic issues in evaluating the properties of thin films. In bulk mechanical structures, the dimensions of structures are made highly accurate by means of machining tools and measurement tools. A mechanically machined structure can be made with a precision of more than one thousandth of its dimensions. However, in MEMS structures, although the absolute error in fabrication is smaller than in mechanically machined structures, the relative accuracies are not good, because the total dimensions are much smaller than the errors. Regarding the thickness of the structure, the deviation is a few percent for most of the deposition methods. In addition, silicon on insulator wafers, whose device layer thickness is determined by the polishing process, has relatively large deviations in the thickness if the device layer is as thin as a few micrometers, because the uniformity of the polishing process is about 0.5–1 µm, irrespective of the total thickness. The lateral dimension, which is mostly determined by photolithography and etching, has the same order of deviation.

Not only the deviation but also the variation of the dimensions of the structure becomes an issue regarding measurement accuracy. Figure 1.2 shows the dimensions of specimens used in published papers on tensile tests of both single-crystal silicon and polysilicon thin films [2–12]. The horizontal and vertical axes represent the length and cross-sectional area of the specimen parallel part, respectively. The plot shows clearly the difference in the dimensions of the thin film specimens. Since the size effect should be considered, direct comparison between the tensile strengths of these specimens is difficult because the dimensions were varied over wide ranges.

1.3.2
Issues Related to Test Apparatus

The differences in the measurement methods become another issue. It is difficult to attribute the differences in measured mechanical properties between measure-

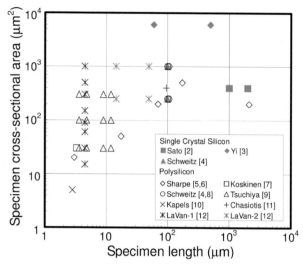

Figure 1.2 Dimensions of previously reported thin-film specimen for uniaxial tensile test. LaVan's specimens [12] are categorized into two types.

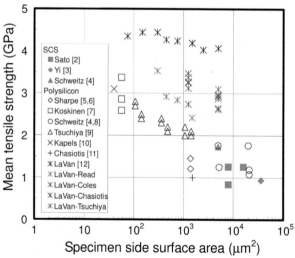

Figure 1.3 Reported tensile strength of both single-crystal silicon and polysilicon films. LaVan's specimens [12] were tested by five institutes.

ment methods. They may include the deviations of all possible parameters, as discussed above. Figure 1.3 shows the measured tensile strength of silicon specimens in the same paper as in Figure 1.2. The average tensile strength was plotted against the side-surface area, which is twice the product of the length and thickness of the specimen parallel part. The plot is based on the result of the size effect analysis of polysilicon specimens, which shows that the fracture origin is located on the side surfaces that are processed by reactive ion etching processes. Brittle materials, whose fracture is dominated by defects contained in the specimen, exhibit a size effect on the strength. The fracture in silicon thin films was often initiated from the side surface of specimens that were formed by dry etching [9]. Therefore, the size effect on strength should be normalized by the side-surface area of the specimens. As can be seen in Figure 1.3, the effect of specimen size appeared in the same experiments and the slope of strength against side-surface area was similar for all experiments. However, the size effect between the different experiments was not clearly observed.

1.3.3
Standards

In methods to evaluate the material properties of bulk materials, international standards are usually established to minimize differences and errors between test machines in measuring properties. Standards on test machines, test specimens and standard specimens to calibrate test machines were established and used to improve the reliability and accuracy of test results. However, standards on the mechanical properties of thin films have not been investigated or established.

The lack of standard methods on the evaluation of thin-film mechanical properties prevents effective material research, as discussed above. It was concluded that many reasons are responsible for the differences between measurements, such as deposition conditions, post-annealing, etching, specimen size effect, deviations in dimensions of specimens, stress concentrations caused by specimen shape and errors resulting from the test apparatus. However, the source of these differences has not been attributed quantitatively and the reliability of each measured value is not confirmed because standard procedures and methods for thin films have not been established.

1.4
Cross-comparison of Thin-film Tensile Testing Methods

To investigate whether differences in tensile strength were caused by the test method, a cross-comparison of existing tensile test methods was carried out. It is difficult to compare test methods from the reported results, as discussed the previous section. Specimens made of the same materials fabricated with the same processes have to be tested in parallel. A round robin test (RRT) scheme was applied to compare the test methods to eliminate the effects of materials, processes, specimen shapes and dimensions. RRTs are evaluations conducted on one specimen at different locations or with different methods to compare the results so that each test method can be checked and evaluated. However, in this case, it is not possible to evaluate the same specimen in the strict sense, since specimens were broken during the fracturing tests. Therefore, the same specimen in this RRT

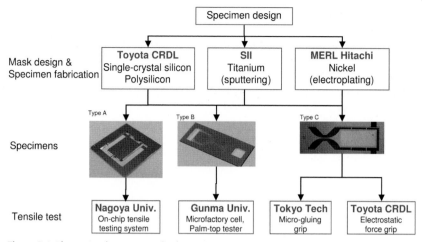

Figure 1.4 Plan to implement round robin test.

was defined as samples fabricated by the same process on a single wafer. Since it is possible to produce multiple microstructures simultaneously with a batch process using silicon micromachining, we can produce samples that are practically the same. We can also minimize variations in the specimen manufacturing process by producing the specimens simultaneously on a single wafer, because the test material undergoes the same fabrication process.

The plan to implement the round robin tests is shown in Figure 1.4. Specimens were designed based on three types of shape conforming to the test methods that will be described below. Three institutes conducted the mask design and fabrication process for each test material. Specimens extracted from a single wafer were distributed to four institutes that conducted tests on these according to the institute's methods.

1.4.1
Tensile Testing Methods

Table 1.1 lists the tensile test methods that were compared in the RRTs. These were characterized by their specimen gripping methods. Tensile stress loading was done by piezoelectric actuators or motorized micrometers. The tensile load was measured with a load cell or the displacement of a double cantilever beam. Elongation in the specimen was measured by gauge mark displacement using image analysis.

Sato et al. at Nagoya University employed an on-chip tensile testing system [2] that integrates a tensile-stress loading system with the specimen chip. The chip converts vertical external load to tensile force on the specimen. From the vertical load and the displacement of the load lever, one can calculate the stress and strain

Table 1.1 Tensile test methods compared in the round robin test.

Institute	Method	Tensile loading	Load (stress) measurement	Strain measurement	Specimen	Ref.
Nagoya University (NU)	On-chip tensile testing system	Motorized micrometer	Double cantilever beam	Image analysis	Type A	2
Gunma University (GU)	Microfactory cell, palm-top tester	Piezo-driven inch-worm	Double cantilever beam	Image analysis	Type B	13
Toyota CRDL (TCRDL)	Electrostatic grip	Piezo-actuator	Load cell	Image analysis	Type C	9
Tokyo Tech	Micro-gluing grip	Magnetostrictive actuator	Load cell	Image analysis	Type C	14

on the specimen by differentiating two measurements of the load–deflection relationship before and after the specimen's fracture.

Saotome et al. at Gunma University [13] used mechanical grip systems applied to thin-film specimens. A cantilever-shaped thin-film specimen was fabricated on a silicon wafer and the free end was fixed to the silicon frames by support beams. After being placed on the tester and fixed by the grip, the specimen was released from the frame by breaking the support beams.

A micro-gluing grip was employed by Higo et al. [14] at Tokyo Institute of Technology (Tokyo Tech), which uses an instant glue to fix the micro-sized specimen. An electrostatic grip was employed by Tsuchiya et al. at Toyota Central Research and Development Laboratories (TCRDL) [9], which uses electrostatic force to chuck the specimen. A cantilever beam with large paddles on its free end was used as the specimen. Electrostatic force for fixing the specimen was generated by applying voltage between the specimen and the chuck device (probe) for conductive materials.

1.4.2
Specimen Design

Three types of RRT specimens were designed. We would have preferred to test specimens with the same shape for the RRTs. However, the specimen shapes and dimensions are completely different from one another because of the test methods. In the five test methods, the size of the specimen chip of the on-chip testing device is 15 mm square, whereas that of the electrostatic force grip is only 1 mm square. It is impossible to use one specimen design for all test methods. Therefore, we used three different designs that had the same length and width over the gauge (parallel) part. We determined the specimen design and dimensions taking the specimen dimensions in Figure 1.1 into consideration. However, the length of the

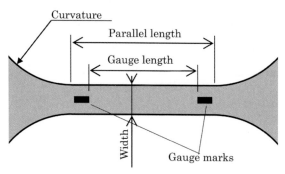

Figure 1.5 Tensile testing part of round robin specimen.

Table 1.2 Dimensions of the specimens for each test method.

Specimen type	A	B	C	
Test method	On-chip	Mechanical	Glue/electrostatic	
Width (μm)	50	50	50	50
Gauge length (μm)	100	500	100	500
Parallel length (μm)	120	600	120	600
Curvature (μm)	500	5000	500	500
No. of specimens	13	14	24	24

three types for specimens could not be designed to be the same. The maximum specimen length for the on-chip test was 100 μm and the minimum length of the mechanical grip was 500 μm. Therefore, the electrostatic/gluing specimen had two types of specimens of different lengths to make comparisons among all methods.

The design of the testing part is shown in Figure 1.5 and the dimensions of each specimen type and the number on one wafer are summarized in Table 1.2. We placed as many of the three types of RRT specimens as possible on a 4-inch wafer, because all the RRT specimens had to be obtained from the one wafer fabricated through the same process.

1.4.3
Materials

Single-crystal silicon, polysilicon, nickel and titanium thin films were selected as the test materials considering their application to micromachines as structural materials. The deposition method and thickness of the test materials are summarized in Table 1.3.

Table 1.3 Thin-film materials evaluated in the round robin test.

Fabrication	Material	Method	t (μm)
TCRDL	SCS SDB	CZ → direct bond	2.0
	Epi	Epitaxial	2.0
	Polysilicon	LPCVD	2.0
Hitachi	Nickel	Electroplating	2.0
SII	Titanium	Sputtering	0.5, 1.0

Silicon is the most frequently used material for micromachine structures because of its superior elastic properties. A single-crystal silicon (SCS) specimen was fabricated from the top layer of a silicon-on-insulator (SOI) wafer. There are various fabrication methods for an SOI wafer: silicon direct bonding (SDB) SOI (SDB-SOI), oxygen ion implantation (SiMOX) and epitaxial growth of silicon {Epi-SOI(CANON ELTRAN [15])). We used SDB–SOI and Epi-SOI because the former is commonly used and the latter has better thickness uniformity (~3%) than SDB-SOI (±0.5 μm). In this chapter, we call these SCS specimens fabricated from SDB-SOI and Epi-SOI SCS/SDB and SCS/Epi, respectively. For the polysilicon specimen, we used a crystallized film from low-pressure chemical vapor-deposited (LPCVD) amorphous silicon using disilane (Si_2H_6) gas as source gas [16].

Electroplated nickel film is also used in LIGA (Lithographie, Galvanoformung, Abformung) processes and other electroplated structures. The test materials deposited on a wafer need to be low tensile stress films and have small thickness variations (<5%). For a nickel specimen, nickel(II) sulfamate tetrahydrate [$Ni(OSO_2NH_2)_2 \cdot 4H_2O$] solution was used as the electrolyte and the current density and the temperature of electroplating were 0.51 A dm^{-2} and 51 °C, respectively, which were optimized for thickness uniformity and internal stress control.

We also selected sputtered titanium films, which are widely used as an electrode material. The titanium film was deposited by sputtering. The argon gas flow rate was controlled and an intermediate cooling process was used for 1-μm thick films to make the internal stress low [17].

1.4.4
Specimen Fabrication

The RRT specimen fabrication process was based on an on-chip tensile testing device. The process involves four lithography and etching steps: specimen shape definition, gauge mark fabrication, silicon wafer etching from the front to define the torsion bar thickness of the on-chip tensile testing device and wafer etching from the back side to release the specimens. The order of these steps differed according to the materials being tested.

The fabrication process for both SCS and polysilicon specimens was similar to the original process, with some steps added for gauge mark fabrication and passivation of the test material and the gauge marks. We used back-side polished SOI and silicon wafers for the single-crystal silicon and polysilicon specimen processes, respectively. For SOI wafers, 1-μm thick SiO_2 film was deposited by plasma-enhanced chemical vapor deposition (PECVD) to passivate the top silicon layer. Then, all wafers were thermally oxidized (0.5 μm) to create a sacrificial layer of the polysilicon specimens and the mask material for anisotropic etching from the back side. Then, LPCVD amorphous silicon film was deposited (2 μm) and annealed in N_2 at 1000 °C to achieve crystallization. This "crystallized" polysilicon film was used as the back-side passivation film and also as the test material. After both single-crystal silicon and polysilicon films had been patterned to the specimen shape with reactive ion etching (RIE), titanium and titanium nitride film was deposited and patterned for the gauge marks. The silicon wafer was anisotropically etched using a tetramethylammonium hydroxide (TMAH) solution from both the front and back sides of the wafer. PECVD SiO_2 films were used to passivate the specimen and gauge mark films. Finally, all oxide layers including sacrificial oxide were removed with buffered hydrofluoric acid (BHF) solution.

Anisotropic silicon etching for nickel specimen fabrication was done first from the front side of the thermally oxidized silicon wafer using potassium hydroxide (KOH) solution. Then a 2-μm thick nickel film was electroplated using sputtered chromium and gold film as a seed layer. The thin gold film was deposited with sputtering and a lift-off process to produce the gauge marks. Then, anisotropic silicon etching from the back side was done. Finally, the thermal oxide film was removed with BHF.

Aluminum films were used in titanium specimen fabrication, as the sacrificial layer and passivation films for TMAH etching, instead of silicon dioxide film, because titanium film is attacked and damaged by the hydrofluoric acid used in silicon dioxide etching. The titanium film was deposited on the aluminum film and a thin gold film for the gauge marks was deposited and patterned. After the titanium film had been patterned to the specimen shape, the aluminum films were deposited again to passivate the specimen. Then, the silicon wafer was anisotropically etched from the back side of the wafer to release the specimen. The etchant was a silicon and ammonium persulfate-dissolved TMAH solution that did not attack the aluminum films [18].

Figure 1.6 shows a processed wafer of a single-crystal-silicon specimen before separating into chips, the three types of single-crystal silicon specimens and the type C specimens for gluing and electrostatic grip of each material.

1.4.5
Results

The results for each material except nickel were obtained from specimens on a single wafer. Nickel specimens were obtained from two wafers. The numbers of specimens tested for each material with each method are different, ranging from 1 to 13.

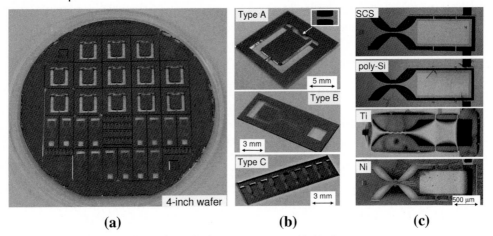

(a) **(b)** **(c)**

Figure 1.6 Fabricated round robin test specimens. (a) 4-inch silicon wafer of single-crystal silicon specimen, just finished the removal of the buried oxide, before cleaving into each chip; (b) three types of the single-crystal silicon specimen chips; (c) type C chip of each material.

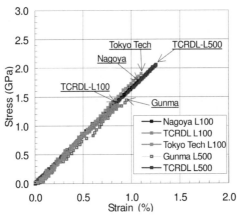

Figure 1.7 Stress–strain curves of single-crystal silicon film fabricated from SDB-SOI wafer. ($w = 20\,\mu m$).

1.4.5.1 Single-crystal Silicon and Polysilicon

Both single-crystal silicon (SCS) films exhibited linear stress–strain relationships and there was good agreement among all the curves obtained by each method, as Figure 1.7 shows for the stress–strain curves of SCS/SDB obtained from each tensile test method. There was little difference in the stress–strain curves for polysilicon and some curves had a non-linear area in the high stress/strain region. The reason for this non-linear part is not clear, but we think that some specimen chucking problems occurred rather than plastic deformation.

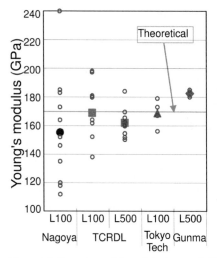

Figure 1.8 Young's modulus of single-crystal silicon from Epi-SOI.

The mechanical properties measured from the stress–strain curves showed good agreement among the test methods. The averages for the Young's modulus of SCS/SDB and SCS/Epi tested by each test method ranged from 135 to 219 and from 155 to 183 GPa, respectively. The averages for polysilicon ranged from 134 to 173 GPa. They ranged around the theoretical modulus. The theoretical Young's modulus of SCS specimens whose tensile axis was in the 110 direction was 168.9 GPa. The polysilicon film was (111) oriented. Young's modulus in the plane direction of (111) oriented polysilicon film did not depend on the in-plane orientation and was 168.9 GPa. The averages of three silicon films measured by TCRDL and Tokyo Tech ranged from 163 to 180 GPa, which well agreed with the theoretical values. The standard deviation for Young's modulus was about 10% of the average Young's modulus. The measured values and averages of Young's modulus of SCS/Epi are plotted in Figure 1.8.

The average tensile strength of SCS/SDB, SCS/Epi and polysilicon ranged from 1.49 to 2.05, from 1.87 to 2.25 and from 1.44 to 2.51 GPa, respectively, and the average fracture strain ranged from 0.92 to 1.20, from 0.87 to 1.45 and 0.96 to 1.52%, which were largest in polysilicon, SCS/Epi and SCS/SDB in that order. This means that polysilicon had the smallest fracture origin. The Weibull plot of each silicon film is shown in Figure 1.9. The Weibull moduli of SCS/SDB, SCS/Epi and polysilicon ranged from 4.7 to 8.6, from 3.1 to 4.8 and from 7.3 to 16.5, respectively, which showed that the deviation in strength of polysilicon was smaller than that in SCS. We can conclude that the deviation in defect size was uniform in the polysilicon specimen. The fracture origin of the silicon specimens was often located on RIE etched surfaces [9]. The etched surfaces of these materials may have had different roughnesses.

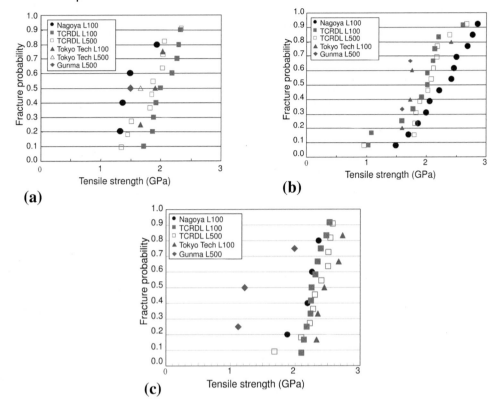

Figure 1.9 Weibull plots for silicon films. (a) Single-crystal silicon from SDB-SOI; (b) single-crystal silicon from Epi-SOI; (c) polysilicon.

1.4.5.2 Nickel

The nickel specimens exhibited brittle fractures with small plastic deformation after their yield point was identified. Fracture surfaces were parallel to the maximum shear stress directions. As shown in Figure 1.10, the stress–strain curves indicated a large difference in both the slope of the curves and the fracture strains between the specimens. The difference in the slopes reflects the difference in Young's modulus. The averages of the Young's modulus, tensile strength and fracture strain ranged from 49 to 185 GPa, from 0.54 to 2.18 GPa and from 0.93 to 2.31%, which showed much larger deviations than those of silicon films. Figure 1.11 shows the measured values and averages of Young's modulus and tensile strength. The largest maximum strains appeared between specimens tested using the mechanical grip and micro-gluing methods, where the loading rate was low. We have to equalize the loading rate in order to compare ductile materials.

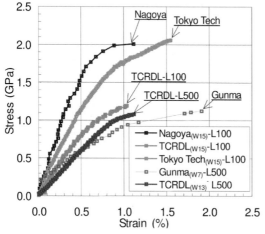

Figure 1.10 Stress–strain curves of electroplated nickel film.

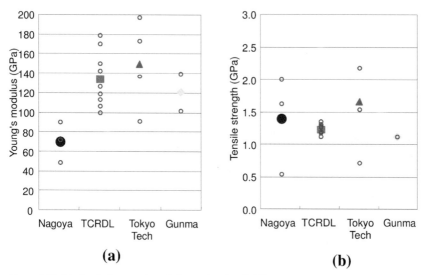

(a)　　　　　　**(b)**

Figure 1.11 Mechanical properties of electroplated nickel film.
(a) Young's modulus; (b) tensile strength.

1.4.5.3 Titanium

The titanium specimens exhibited brittle fractures with plastic deformation after their yield point was identified as shown in the stress–strain curves in Figure 1.12. In contrast to the nickel films, there are small differences between test methods. The Young's modulus and tensile strength are plotted in Figure 1.13. The average Young's modulus was about 100 GPa, which is smaller than that of bulk titanium of 115 GPa [19]. The deviation in modulus was caused by the deviations in dimen-

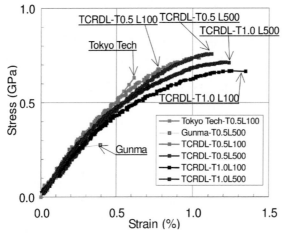

Figure 1.12 Stress–strain curves of sputtered titanium film.

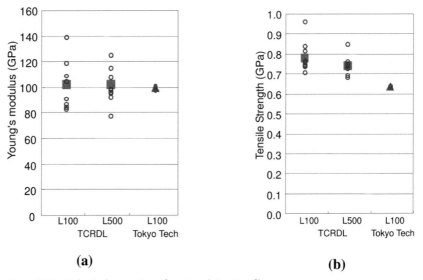

(a)

(b)

Figure 1.13 Mechanical properties of sputtered titanium film.
(a) Young's modulus; (b) tensile strength.

sions, especially specimen thickness. The average tensile strength ranged from 0.64 to 0.78 GPa. The deviation in strength was small. Some specimens had very large (>10%) maximum strains. In these specimens, large slips along the maximum shear stress directions appeared. However, Ogawa et al. reported brittle fractures and small maximum elongations in tensile tests of sputtered titanium films [20].

This difference in fracture behavior may be caused by the deposition conditions [17].

Not all of the stress–strain curves of each test method for titanium films could be obtained because there were some problems, e.g. the specimens were too thin to calculate the applied load for on-chip tensile test methods or the specimens were damaged during the specimen chucking procedures used in mechanical grip methods.

1.4.6
Discussion

The RRT results revealed that there were no apparent differences between measuring methods and the measured properties and their deviations had almost the same values. Figure 1.9 shows the Weibull plots of the silicon specimens. The plotted points in each graph represent the strength of the specimen from one wafer. The slope of each plot is similar for the same materials. This means that the deviation in strength, i.e. the deviation in the size of the fracture origin, is the same, which means that specimens tested with all tensile test methods fractured in the same fracture mode. These results confirm the accuracy and repeatability of all these methods.

The standard deviation of the Young's modulus of silicon films ranged from 5 to 20%, which was larger than the estimated deviation in specimen dimensions. We have to identify the source of these deviations in order to reduce them.

To compare the measured Young's modulus from each test method statistically, the hypothesis that "there is no difference between the two test methods in measuring Young's modulus" was tested using a *t*-test at significance level of 5%. The hypothesis was accepted in most of the comparisons, but rejected in comparison with the on-chip tensile test method with the other methods. There are two possible reasons for the difference with the on-chip tensile test method. First, the mechanism that converts pushing force to tensile force would lead to errors in the force measurement. The rotation axis of the torsion bar would change during the testing. Second, the differential measurement would cause errors, because the difference in the Young's modulus was larger when the cross-section of the specimen was smaller. In such a case, the torsion bar stiffness is much larger than the specimen stiffness.

The mechanical grip methods had smaller values for strength measurement, which may have been caused by the effect of the size or stress concentration on the rounded part of the specimens that had a larger radius of curvature.

In summary, there were no systematic differences between the test techniques compared in this experiment. One can choose any method in the evaluation of the strength of thin films for his object, process compatibility or equipment. However, there is a weakness in each method. For example, the electrostatic grip has a limit of the maximum tensile force that can be applied to the specimen because of the gripping force limit. The on-chip tensile testing device has a limit of the maximum specimen length. The displacement applicable at the end of the specimen is limited by the torsion bar strength.

1.5
International Standards on MEMS Materials

The measurement of thin-film mechanical properties is crucial for the design and evaluation of MEMS devices. A lot of research has been carried out to evaluate the repeatability, accuracy and data reliability of various measurement methods for thin-film mechanical properties, because the information from these studies so far is not sufficient. The development of international standards on MEMS materials and their properties measurement methods will solve these problems. There had been an understanding that the MEMS industry does not require standards because of the variations in fabrication methods and dimensions. Recently, the development of standards in the MEMS field has been started in order to establish the fundamentals of reliability evaluations, especially on material properties. In this section, recent activities in the development of standards in MEMS are described. In addition, published international standards and currently active projects on measurements of mechanical properties are introduced.

1.5.1
MEMS Standardization Activities

1.5.1.1 IEC

The International Electrotechnical Commission (IEC) is the international standardization association for the electrical and electronics industries. The Technical Committee on Semiconductor Devices (TC47) has a working group on microelectromechanical systems (TC47/WG4). This working group is the only committee in the MEMS field in the international standardizing association. The group has published three international standards and is discussing three projects for new proposals on MEMS standards. The first published standard is on terminology (IEC62047-1), published in 2005. It contains more than 100 terms with definitions. The second and third one were published in 2006 on tensile test methods for thin films. Currently under discussion are standards on general definitions, RF-MEMS switches and fatigue test methods. Japan and Korea are active in the working group and each has a plan to propose new standard items in the future.

1.5.1.2 ASTM International

ASTM International is an international standards organization that develops and produces technical standards mainly for materials. The Technical Committee on Fatigue and Fracture (E08) has a Task Group (E08.05.03) on Structural Films and Electronic Materials that develops standards for electronic and micromechanical applications. Standards on in-plane length, residual strain and strain gradient measurement of thin films using an optical interferometer (E2244-06, E2245-06 and E2246-06) have been published. They also carried out a round robin test on thin-film tensile testing [12].

1.5.1.3 SEMI

SEMI is an industry association mainly for semiconductor manufacturing supply. It is also working on standards related to semiconductor fabrication, such as the specification of wafers and process gases, an evaluation method for process techniques and process management. SEMI has an interest in MEMS devices due to the compatibility of their fabrication processes. Standardization in the MEMS fabrication area is an active consideration. Three standards have been published: Guide to Specifying Wafer–Wafer Bonding Alignment Targets (MS1-0307), Test Method for Step-height Measurements of Thin, Reflecting Films Using an Optical Interferometer (MS2-0307) and Terminology for MEMS Technology (MS3-0307). Development work on wafer bond strength test methods is in progress.

1.5.1.4 Micromachine Center in Japan

The Micromachine Center (MMC) is a nonprofit foundation dedicated to supporting the establishment of a technological basis of future micromachines/MEMS, and to support the development of MEMS/micromachine industries. MMC is dedicated to standardization in this field and has published technical reports on terminology and evaluation methods for material properties. Based on these studied and projects on standardization research supported by the government, they have proposed standards on terminology, tensile test methods and fatigue test methods to IEC TC47/WG4.

1.5.2
International Standards on Thin-film Uniaxial Stress Testing

The IEC standard on tensile test methods for thin-film materials (IEC 62047-2) specifies tensile test methods for thin-film materials whose thickness is less than $10\,\mu m$ and lateral dimensions (length and width) are less than 1 mm. This standard is a guide for the repeatable, reliable and accurate tensile testing of thin films, in which the specifications of specimen, test machine and test conditions are identified, but a specific test method is not required. In the Section 4 on test methods and test apparatus, specimen gripping is pointed out as an important specification for test apparatus. Some gripping methods are explained in an appendix. In Section 5 on specimens, the specification of the specimen is described, which considers dimension errors and shape errors of microfabricated specimens.

The IEC standard on thin-film standard test pieces for tensile testing (IEC 62047-3) specifies a standard test piece for thin-film tensile testing in order to qualify the accuracy and repeatability of a tensile testing machine. This standard recommends using single-crystal silicon as a standard specimen.

These two standards related to tensile testing of thin films were compiled from the results of cross-comparisons among tensile test methods described in the previous section. Currently, a new standard on axial fatigue test methods for thin-film materials is being discussed, which is based on the international standard on fatigue test methods for metallic materials (ISO 1099) and referred to IEC 62047-2. The differences between bulk and thin films are specified.

1.6
Conclusion

In this chapter, the contribution of the mechanical properties of thin films to the properties of MEMS devices, especially reliability, has been described. Not only the strength and fatigue, but also the elastic properties are important for MEMS reliability. There are many difficulties in their evaluation, which mainly come from their small dimensions. The cross-comparison of evaluation methods to solve these difficulties and international standards compiled from the results of such comparisons are described.

In order to realize highly functional and highly reliable MEMS devices, accurate, repeatable properties should be measured, in which the measured properties can be compared with each other. These activities should extend to various evaluations in MEMS research and development.

References

1 L. Hornbeck, Digital light processing and MEMS: timely convergence for a bright future, *Proc. SPIE* **2639** (1995) 2.

2 K. Sato, T. Yoshioka, T. Ando, M. Shikida, T. Kawabata, Tensile testing of silicon film having different crystallographic orientations carried out on a silicon chip, *Sens. Actuators A* **70** (1998) 148–152.

3 T. Yi, L. Li, C. J. Kim, Microscale material testing of single crystalline silicon: process effects on surface morphology and tensile strength, *Sens. Actuators A* **83** (2000) 172–178.

4 J. Å. Schweitz, F. Ericson, Evaluation of mechanical materials properties by means of surface micromachined structure, *Sens. Actuators A* **74** (1999) 126–133.

5 W. N. Sharpe, B. Yuan, R. Vaidyanathan, R. L. Edwards, New test structures and techniques for measurement of MEMS Materials, *Proc. SPIE* **2880** (1996) 78–91.

6 W. N. Sharpe, K. T. Turner, R. L. Edwards, Tensile testing of polysilicon, *Exp. Mech.* **39** (1999) 162–170.

7 J. Koskinen, J. E. Steinwall, R. Soave, H. H. Johnson, Microtensile testing of free-standing polysilicon fibers of various grain sizes, *J. Micromech. Microeng.* **3** (1993) 13–17.

8 S. Greek, F. Ericson, S. Johansson, M. Fürtsch, A. Rump, Mechanical characterization of thick polysilicon films: Young's modulus and fracture strength evaluated with microstructures, *J. Micromech. Microeng.* **9** (1999) 245–251.

9 T. Tsuchiya, O. Tabata, J. Sakata, Y. Taga, Specimen size effect on tensile strength of surface micromachined polycrystalline silicon thin films, *J. Micro ElectroMech. Syst.* **7** (1998) 106–113.

10 H. Kapels, R. Aigner, J. Binder, Fracture strength and fatigue of polysilicon determined by a novel thermal actuator, *IEEE Trans. Electron Devices*, **47** (2000) 1522–1528.

11 I. Chasiotis, W. G. Knauss, Microtensile tests with the aid of probe microscopy for the study of MEMS materials, *Proc. SPIE* **4175** (2000) 92–102.

12 D. A. LaVan, T. Tsuchiya, G. Coles, W. G. Knauss, I. Chasotis, D. Read, Cross

comparison of direct strength testing techniques on polysilicon films, in *Mechanical Properties of Structural Films*, ASTM STP 1413, ASTM, Philadelphia, 16–26, 2001.

13 Y. Saotome, Y. Nakazawa, S. Kinuta, Microfactory cells for *in-situ* micromaterials and microstructures testing, in Proceedings of the 2nd International Workshop on Microfact., Fribourg, Switzerland, October 9–10, 2000, 99–102, 2000.

14 Y. Higo, K. Takashima, M. Shimojo, S. Sugiura, B. Pfister, M. V. Swain, Fatigue testing machine of micro-sized specimens for MEMS applications, *Mater. Res. Soc. Symp. Proc.* **605** (2000) 241–246.

15 T. Yonehara, K. Sakaguchi, ELTRAN®; novel SOI wafer technology, *JSAP Int.* **4** (2001) 11–16.

16 H. Funabashi, T. Tsuchiya, Y. Kageyama, J. Sakata, Fabrication technology of three layer polysilicon microstructures without CMP for gyroscope, in *Technical Digest of the 10th International Conference of Solid-State Sensors and Actuators (Transducers 99), Sendai, Japan, June 7–10, 1999,* 336–339, 1999.

17 T. Tsuchiya, M. Hirata, N. Chiba, Young's modulus, fracture strain and tensile strength of sputtered titanium thin films, Thin Solid Films **484** (2005) 245–250.

18 G. Yan, P. C. H. Chan, I.-M. Hsing, R. K. Sharma, J. K. O. Sin, Y. Wang, An improved TMAH Si-etching solution without attacking exposed aluminum, *Sens. Actuators A* **89** (2001) 135–141.

19 M. Chinmulgund, R. B. Inturi, J. A. Barnard, Effect of Ar gas pressure on growth, structure and mechanical properties of sputtered Ti, Al, TiAl and Ti3Al films, *Thin Solid Films* **270** (1995) 260–263.

20 H. Ogawa, K. Suzuki, S. Kaneko, Y. Nakano, Y. Ishikawa, T. Kitahara, Tensile testing of microfabricated thin films, Microsystem Technologies, **3** (1997) 117–121.

2

Elastoplastic Indentation Contact Mechanics of Homogeneous Materials and Coating–Substrate Systems

Mototsugu Sakai, Department of Materials Science, Toyohashi University of Technology, Tempaku-cho, Toyohashi 441-8580, Japan

Abstract

Micro/nanoindentation technique is the most appropriate for the mechanical characterization of materials and structures in volumes with extremely small dimensions such as MEMS, optoelectronic devices, so forth. The physical significances are intensively addressed of the contact mechanics parameters obtained in instrumented micro/nanoindentation test systems applied to bulk homogeneous as well as coating/substrate composite materials in elastic/elastoplastic regimes. The details of the experimental procedures and analyses for characterizing the contact parameters including the Meyer hardness and the elastic modulus are also discussed.

Keywords

indentation; elastoplastic; elastic; contact mechanics; homogeneous materials; coatings

Reliability of MEMS.
Edited by O. Tabata, T. Tsuchiya
Copyright © 2013 WILEY-VCH Verlag GmbH & Co. KGaA, Weinheim
ISBN: 978-3-527-33501-5

2.1
Introduction

In 1881, Hertz, just 24 years old at the time, published his classical paper, *On the contact of elastic solids* [1]. His major interest was in the optical interference between glass lenses. His theoretical considerations on the contact pressure distributions of solid spheres, on the basis of analogy with electrostatic potential theory, enabled the significant influence on the interference fringe patterns at contact to be understood. The classical approach to finding the elastic stresses and their distributions due to the surface tractions was also made by Boussinesq in 1885 with the use of the theory of potential [2, 3]. This approach was then extended by Love in the 1930s [4] and also by Sneddon in 1965 for axisymmetric elastic indentation problems [5].

With purely elastic materials that have small elastic moduli, such as rubber-like materials, the elastic properties play a dominant role in contact deformation. With ductile metals, however, due to their large elastic moduli, the contact deformation is predominantly out of the elastic regime, often involving highly plastic deformation and flow and leaving a finite residual impression after unload. In 1908, based on experimental results on the geometric similarity in spherical indentation, Meyer first proposed the concept of indentation contact hardness (Meyer hardness) as a measure of plastic deformation [6]. He defined the hardness as the ratio of the contact load to the *projected* area of the residual impression of contact. If the materials are ductile enough, the Meyer hardness H_M is related to the plastic yield strength Y through a constraint factor C as $H_M = CY$. Prandtl (1920) examined the plastic flows beneath a flat punch [7] and then Hill (1950) extended the issues into what is now well known as slip line field theory [8]. They predicted the C value to be 2.57. The constraint factor C, in fact, varies from about 2.5 to 3.2, depending on the geometry of the contacting indenter and also on the frictional condition at the interface between the indenter and the material indented [9, 10].

Staudinger proposed a novel concept of macromolecules in 1920 [11] and of their molecular network structures in 1929 [12], followed by the pioneering work of Carothers in 1929 on artificially synthesized organic polymers [13]. Along with a series of outstanding studies of polymer physics and rheology conducted by Flory in the 1940s [14] and by Ferry in the 1950s [15], an upsurge of interest in the science and engineering of synthesized organic polymers enhanced studies on time-dependent viscoelastic deformations and flows (polymer rheology). With this historical background, Radok first discussed in 1957 the viscoelastic indentation problem of a sphere pressed into contact with a linear viscoelastic body [16]. He suggested a simple approach to time-dependent indentation for finding a viscoelastic solution in cases where the corresponding solution for a purely elastic body is known from the use of the fact that the Laplace transform of the viscoelastic constitutive equation results in a purely elastic equation. His theory was then more generalized by several researchers [17–20]. Due to the experimental difficulties associated with time-dependent contact deformation and flow, the experimental scrutiny of the theories and attempts to apply them to the experimental characterization of viscoelastic materials were not made until the end of 1990s [21–26].

The earliest experiments were conducted by Tabor, in which measurements of the relationship between indentation load P and penetration depth h were made when a spherical indenter was pressed into contact with metals [9]. The elastic recovery of a hardness impression during spherical indentation unloading was then examined in 1961 by Stilwell and Tabor in terms of the elastic modulus of metallic materials indented [27] and also by Lawn and Howes for pyramidal indentations of various brittle ceramics and ductile metals [28]. In 1981, Shorshorov et al. first considered the experimental determination of the Young's modulus of a material indented by the use of the stiffness of the indentation load P versus penetration depth h unloading curve and suggested the importance of constructing an instrumented indentation apparatus [29]. An attempt was then made by Newey et al. in 1982 to construct an instrumented micro/nanoindenter to obtain a continuous record of the penetration depth h in the micro/nanoregime as a function of applied load and time and to apply it to study the hardness of ion-implanted layers [30]. Along with the dramatic progresses and developments in electronic sensor devices with high precision that started in the mid-1980s for measuring displacement on the nanoscale and the load in micronewtons, several types of instrumented micro/nanoindentation apparatus were developed and applied to studies on the elastoplastic surface deformation of bulk materials and thin coatings [31–57]. Through a number of these intensive and extensive experimental studies, it has been well recognized that such micro/nanoindentation apparatus is efficient in particular for studying the characteristic mechanical properties of thin coatings, micro/nanocomposites, advanced engineering components having micro/nanoscales constructed in microscale electromechanical systems (MEMS) and so forth.

2.2
Instrumented Micro-/Nanoindentation Apparatus

The schematics of instrumented micro/nanoindentation test systems with different depth-sensing configurations are depicted in Figure 2.1 [51]. The displacement gauge is mounted on the test frame and senses the motion of the load train relative to the test frame (see Figure 2.1a). This type of indentation apparatus includes most commercially available test systems. In Figure 2.1b, the displacement gauge is mounted on the holder/stage of the test specimen and senses the *relative penetration* of the load train immediately above the indenter tip. The observed penetration depth h_{obs} in the former system (Figure 2.1a), referred to as a "test-frame-based system", inevitably and significantly includes undesirable contributions from the elastic compliance of test frame/load train and also the contact clearances of the test holder/stage. This system, however, has an advantage for easily monitoring the indentation behavior at elevated temperatures, in hostile environments, and so forth. The observed indentation displacement h_{obs} is related to the true

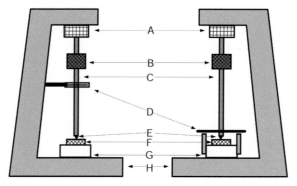

(a) Test-frame-based measurement of penetration depth   (b) Test-stage-based measurement of penetration depth

Figure 2.1 (a) Test frame-based and (b) test stage-based instrumented indentation test systems. A, actuator; B, load detecting sensor; C, load train; D, displacement (depth) detecting sensor; E, indenter; F, test specimen; G, specimen holder; H, test frame.

penetration depths h by the following expression in terms of the undesirable displacement Δh resulting from the elastic deformations of the test frame and load train coupled with mechanical contact clearances:

$$h_{\text{obs}} = h + \Delta h \tag{1}$$

A precise calibration for Δh must be made in the experiment prior to indentation testing. The direct measurement of Δh can be made by the use of a diamond flat punch indenter (with a diameter exceeding 100 μm that is equivalent to its stiffness of about $1 \times 10^8\,\mathrm{N\,m^{-1}}$) pressed into contact with a very stiff specimen, such as a block of sintered SiC ceramic, where the observed "apparent" penetration h_{obs} is equivalent to Δh due to the negligibly small value of the true penetration h. The $P - \Delta h$ calibration curve thus obtained in the instrumented indentation apparatus is then subtracted from the conventional $P - h_{\text{obs}}$ curve, resulting in the "true" $P - h$ curve.

When the *elastic* test-frame compliance is the major contribution to Δh, there is the following approximation for an axisymmetric indentation [41]:

$$C_{\text{obs}} = C_{\text{f}} + \frac{\sqrt{\pi}}{2E'_{\text{app}}} \frac{1}{\sqrt{A_{\text{C}}}} \tag{2}$$

where C_{obs} and C_{f} are the elastic compliances of the observed test system and the test frame, respectively. The following relationships must be noted: $h_{\text{obs}} = C_{\text{obs}}P$ and $\Delta h = C_{\text{f}}P$. In Eq. (2), A_{c} and E'_{app} are, respectively, the contact area of indenter and the apparent elastic modulus of the system, defined by $(E'_{\text{app}})^{-1} = (E')^{-1} + (E'_{\text{i}})^{-1}$ in terms of the elastic modulus E' [$E' = E/(1 - v^2)$; E is the Young's modulus and

ν is Poisson's ratio] of a standard test specimen used for C_f calibration (a fused-silica glass with $E' = 74.1\,GPa$ is widely used as the standard material) and the elastic modulus of indenter E_i' (the value is $1147\,GPa$ for diamond indenter) [41]. Accordingly, a plot of C_{obs} versus $1/\sqrt{A_c}$ is linear for a given test material and the intercept on the y-axis of the plot is a direct measurement of the test frame compliance C_f. The best value of C_f can be obtained for larger indentations giving larger A_c values, where the second term on the right-hand side of Eq. (2) becomes small. The experimental estimate of A_c is made using the unloading stiffness S {the slope dP/dh of unloading $P - h$ line at the maximum load using the Oliver–Pharr approximation $S = (2/\sqrt{\pi})E_{app}'\sqrt{A_c}$ [41]}, if the indentation behavior is brittle enough. Some further considerations on the iterative determination of C_f and A_c are given later in this section and also in Section 2.4.4.

The test stage-based system shown in Figure 2.1b, in contrast to the test frame-based system in Figure 2.1a, has been designed to minimize or even eliminate the undesirable contribution of Δh. The displacement measured in this test system is only that of the penetration depth of the test specimen h with a very minor elastic contribution of the diamond indenter and its mount, provided that the contact clearance between the specimen and its holder is carefully negated [36, 38, 42, 46]. The instrumented indentation test system designed for minimizing or eliminating the Δh contribution to h_{obs} is critically important, as the empirical procedures and calibrations that are inevitably required in the test frame-based system include several unacceptable uncertainties associated with irreversible contact clearances at mechanical joints along with the approximations included in Eq. (2) due to the assumptions included in estimating A_c.

A piezo or an electromagnetic actuator is widely utilized as the loading actuator. A load sensor with a precision of about $\pm 1\,\mu N$ and a displacement sensor with a precision of about $\pm 0.1\,nm$ are commercially available; a strain gauge-mounted load cell for the former and a linear transducer or a capacitive displacement gauge for the latter are, in general, mounted on conventional micro/nanoindentation test systems.

The geometric configurations are shown in Figure 2.2 of the standard Vickers (tetragonal pyramid) and Berkovich (trigonal pyramid) indenters. Due to the trigonal pyramid configuration, with the Berkovich geometry it is easier to fabricate a sharp tip (a tip radius of <150 nm) than with the tetragonal Vickers indenter. This is the main reason why the Berkovich indenter is more widely utilized in instrumented nanoindentation test systems.

The geometric parameters of the Vickers indenter are shown in Figure 2.2a [51]. The apex angle (the diagonal face-to-face angle 2α) is designed to be $136°$ since this is the angle subtended by the tangents of a Brinell sphere when the ratio of the contact diameter (d) to the diameter of Brinell sphere (D) is 0.375, the most recommended ratio for Brinell hardness testing. The inclined face angle β is, therefore $22.0°$ and the diagonal edge-to-edge angle 2ψ is $148.1°$ with the relation of $\beta = \sqrt{2}\cot\psi$. The projected contact area A_c at the contact depth h_c is, due to the geometric similitude, easily related to the contact depth h_c by $A_c = gh_c^2$ with the geometric factor $g = 24.5$ ($\equiv 4/\tan^2\beta$).

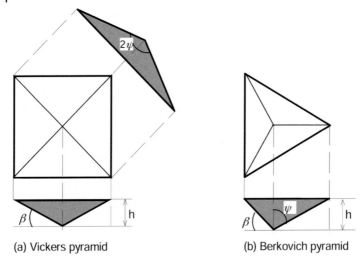

(a) Vickers pyramid   (b) Berkovich pyramid

Figure 2.2 Tip geometries of (a) Vickers pyramid and (b) Berkovich pyramid indenters.

The Berkovich indenter is designed to provide the same contact area to penetration depth relationship as the Vickers indenter; the inclined face angle β is 24.7°, the apex angle (included angle) ψ is 77.1° and the projected edge length $a = (2/\tan \beta)h$ (= 4.35h), resulting in $g = 24.5 (\equiv 3\sqrt{3}/\tan^2 \beta)$ [51].

Upon machining a pyramidal diamond indenter, the tip in reality is not ideally sharp, being truncated or rounded to several tens of nanometers. This geometric uncertainty associated with machining yields a critical difficulty in nanoindentation tests for penetration depths less than about 100 nm. An experimental calibration for the tip truncation/roundness is recommended prior to hardness testing. For a pyramid indenter with a truncated tip, the contact area is expressed by the contact depth h_c in terms of the tip truncation Δh_t as follows [58]:

$$A_c = gh_c (h_c + 2\Delta h_t) \tag{3}$$

or, in an alternative way, using an empirical equation [41]:

$$A_c = gh_c^2 + a_1 h_c + a_2 h_c^{1/2} + a_3 h_c^{1/4} + \ldots + a_8 h_c^{1/128} \tag{4}$$

where a_i (i = 1, 2, 3, . . . , 8) are constants determined by curve-fitting procedures. Applying Eq. (3) or (4) to Eq. (2) along with the experimental measurement of the compliance values C_{obs} for various values of penetration depths h in the *purely elastic regime*, a simultaneous determination of the frame compliance C_f and the tip truncation Δh_t can be conducted in an iterative manner [41, 58]. It is easily confirmed by examining the reversible loading–unloading $P - h$ relationship (that is, the unloading $P - h$ line faithfully coincides with the reloading line in elastic indentation) whether the pyramidal indentation is elastic or not.

A conical indenter (included apex angle of about 120°) with a rounded tip is also widely utilized in micro/nanoindentation testing. Tip radii R ranging from about 2 to 10 μm are generally favored for the nanoindentation test regime. The rounded-tip indenter thus fabricated can be approximated to a spherical indenter having the same radius R, allowing the experimental determination of the elastic modulus (Young's modulus) of the material indented in its elastic regime using the following Hertzian contact equation [1, 3, 5, 10]:

$$P = \frac{4}{3} E'_{app} \sqrt{R} h^{3/2} \tag{5}$$

or its alternative expression in terms of the mean contact pressure $p_m (= \pi a^2)$ (a is the contact radius):

$$\frac{P}{\pi R h} (\equiv p_m) = \frac{4}{3} \frac{E'_{app}}{\pi} \left(\frac{h}{R}\right)^{1/2} \tag{6}$$

where use has been made of the approximation $a^2 \approx Rh$ in Hertzian elastic contact for $a \ll R$ [10]. Accordingly, using the observed $P - h$ elastic curve and the tip radius R in the P/Rh versus $(h/R)^{1/2}$ linear plot, we can determine the elastic modulus E'_{app} from the slope of the linear plot and then E' of the test material through the relation $(E'_{app})^{-1} = (E')^{-1} + (E'_i)^{-1}$.

2.3
Indentation load *P* Versus Penetration Depth *h* Relationship

Figure 2.3 represents schematically the three specific contact behaviors in their $P - h$ loading–unloading hysteresis for materials having time/rate-independent mechanical properties: (a) purely plastic, (b) elastoplastic and (c) purely elastic

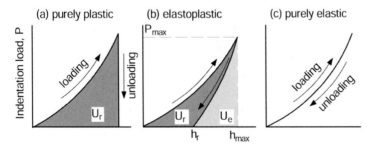

Penetration depth, h

Figure 2.3 Load *P* vs. penetration depth *h* hysteresis of an indentation loading–unloading cycle for (a) purely plastic, (b) elastoplastic and (c) purely elastic materials.

materials [42]. Most "ductile metals" exhibit contact deformation similar to Figure 2.3a, where the elastic recovery of the hardness impression associated with the unloading process is negligibly small relative to the penetration depth, meaning that the residual depth after complete unloading is equivalent to the depth at the maximum penetration. The hysteresis loop energy U_r, which is defined as the area enclosed by the $P - h$ loading–unloading paths, is the energy consumption associated with creating the hardness residual impression [42].

In contrast to Figure 2.3a, as shown in Figure 2.3c, the unloading path of purely elastic materials goes faithfully back along its loading path to the origin of the $P - h$ diagram, leaving none of residual impression after complete unloading. This means that none of the irreversible energies are consumed in the loading–unloading cycle (that is, $U_r = 0$). Soft organic rubbers and brittle glassy carbon [59] are included among these elastic materials showing this purely elastic contact deformation.

Typical brittle and hard polycrystalline materials such as ordinary sintered ceramic materials, single-crystalline materials including metallic silicon and some hard metals will follow Figure 2.3b, where the penetration depth at maximum load shows significant elastic recovery, yet leaving a finite residual depth h_r after unloading. It must be noted in the conventional Vicker/Berkovich pyramid and also spherical indentation tests for pure elastic (Figure 2.3c) and elastoplastic (Figure 2.3b) surface deformations that *the in-surface elastic recovery is negligibly small*, even if the in-depth elastic recovery is significant. In other words, the projected contact area A_c at the maximum penetration is approximately equal to the projected area A_r of the residual impression after complete unloading, that is, $A_r \cong A_c$ [53, 60, 61].

The $P - h$ loading and unloading lines are given, respectively, by the following general equations [51, 53, 55, 62]:

$$P = k_1 h^n \tag{7}$$

and

$$P = \alpha(h - h_r)^m \tag{8}$$

where the frontal coefficients, k_1 and α, are constants, being referred to as the loading and unloading parameters, respectively. The loading exponent (n) is 1.0, 1.5 (= 3/2) and 2.0 for a flat-ended punch, sphere and cone (pyramid) indentations, respectively. The value of the unloading exponent (m) is basically equal to the corresponding loading n value only when the surface profile of residual hardness impression faithfully represents the geometry of the indenter, i.e. only when the impression cavity is ideally spherical although the radius R of impression is always larger than that of the spherical indenter pressed into contact and only when the residual hardness impression is ideally conical (pyramidal) although its inclined face angle β (see Figure 2.2) is always smaller than that of the conical (pyramidal) indenter due to the in-depth elastic recovery [55]. However, in general, the impression profile is not faithfully spherical or conical (pyramidal). In pyramidal indentation, by way of example, the side surfaces of residual impression are always somewhat convex (not being composed from inclined "flat" planes). In such cases, there is a finite difference between the m and n values: m is always smaller than n when the impression surface is convex [55].

2.4
Physics of Elastoplastic Contact Deformations in Conical/Pyramidal Indentation
[42, 51, 53, 55, 60–63]

A conical/pyramidal indentation contact accommodates the elastic, elastoplastic and/or viscoelastic surface deformation(s) over the contact region to avert the stress singularity induced by the acute tip. The contact stresses are highly concentrated beneath the tip and decrease significantly in intensity with distance from the contact region. Meyer's principle of geometric similarity for conical/pyramidal indentation contact suggests that the presence of a representative stress $\bar{\sigma}$ and a representative strain $\bar{\varepsilon}$ properly representing the mechanical field at a representative location in the subsurface [62]; the geometrically similar indentations produce similar stress/strain fields around the contact region. The mean contact pressure, defined by the ratio of indentation load P to the projected contact area A_c at a contact depth of penetration h_c, has been utilized as the representative stress $\bar{\sigma}$:

$$\bar{\sigma} = \frac{P}{A_c} \tag{9}$$

As the representative strain for conical/pyramidal indentation contact, the following expression may be defined in its differential form:

$$d\bar{\varepsilon} = c\frac{dh}{h} \tag{10}$$

where the differential penetration depth is given by dh. The frontal coefficient c in Eq. (10) is a constant that is dependent on the tip geometry of indenter.

2.4.1
Elastic Contact

Consider a cone/pyramidal indenter pressed into contact with an ideally elastic half-space obeying the Hookean constitutive equation

$$\bar{\sigma} = E'\bar{\varepsilon} \tag{11}$$

where E' is the plane strain elastic modulus defined by $E/(1 - v^2)$ in terms of the Young's modulus E and Poisson's ratio v. Substituting Eqs. (9) and (10) into Eq. (11) leads to a simple and straightforward derivation of Sneddon's elastic solution for a conical/pyramidal indenter with inclined face angle β (see Figure 2.2):

$$P = k_e h^2 \tag{12}$$

where

$$k_e = \frac{E' \tan \beta}{2} \left(\frac{g}{\gamma_e^2} \right) \tag{13}$$

Use has been made of the relation $c = \tan \beta$ in Eq. (10) to give Eq. (13) conforming to Sneddon's solution. The g factor in Eq. (13) characterizes the geometric similarity of the cone/pyramid indentation through the relationship $A_c = gh_c^2$ (see Section 2.2). The surface profile of hardness impression at a total penetration depth h is described by γ_e as $h = \gamma_e h_c$ using the contact depth h_c. The γ_e value for conical indentation is $\pi/2$ (Love's solution) [4, 5, 10], meaning that the elastic impression is always sinking-in (the impression profile is sinking-in for $\gamma_e > 1.0$ and piling-up for $\gamma_e < 1.0$).

2.4.2
Plastic Contact

The plastic flow of an elastoplastic material can be described by the yield strength Y. When the applied stress σ exceeds Y, the material exhibits an irreversible plastic flow. Consider a conical/pyramidal indenter pressed into contact with a *perfectly plastic half-space* having a yield strength Y. In perfectly plastic indentation, the representative stress $\bar{\sigma}$ given in Eq. (9) defines the "true hardness", H, as a size- and geometry-invariant characteristic material parameter for "plasticity" [42, 60, 62]:

$$\bar{\sigma} \left(\equiv \frac{P}{A_c} \right) = H \tag{14}$$

The true hardness H defined in Eq. (14) for a perfectly plastic material is then related to the yield strength Y by the following Marsh relationship:

$$H = CY \tag{15}$$

where the constraint factor C is about 3.0, ranging from 2.5 to 3.2 depending on the indenter's geometry and the contact friction [60]. Applying the relationships $A_c = gh_c^2$ and $h = \gamma_p h_c$ to Eq. (14) results in the following $P - h$ relation for purely plastic indentation contact:

$$P = k_p h^2 \tag{16}$$

where

$$k_p = H \left(\frac{g}{\gamma_p^2} \right) \tag{17}$$

The surface profile of impression is, in general, piling-up for purely plastic indentation, being well characterized by γ_p (<1.0) in a quantitative manner.

2.4.3
Elastoplastic Contact [42, 51, 60, 62]

The preceding considerations indicate that the $P - h$ relationships both for purely elastic and purely plastic bodies are expressed by the quadratic relationships in Eqs. (12) and (16) for conical/pyramidal indentation. This suggests that the $P - h$ relation of an elastoplastic half-space may also be quadratic. These quadratic equations basically result from the quadratic nature (geometrically quadratic similarity) of the depth–area relation $A_c = g h_c^2$ for conical/pyramidal indentation.

A simple elastoplastic model and its indentation contact mechanics are demonstrated in what follows to gain a deep physical insight into the elastoplastic indentation parameters, in particular the impression hardness (Meyer hardness). Consider a simple model comprising a purely elastic component having an elastic modulus E' and a purely plastic component having the yield strength Y and so having a true hardness H ($\approx CY$) that is connected in series in the sense of the Maxwell combination. The nominal indentation load P and the total penetration depth h of this elastoplastic model are then expressed in terms of the loads (P_e, P_p) and the depths (h_e, h_p) of the respective elastic (e) and plastic (p) component materials such as $P = P_e = P_p$ and $h = h_e + h_p$ due to their serial combination. Substituting Eqs. (12), (13), (16) and (17) into these constitutive relations of the elastoplastic model, we obtain the following quadratic $P - h$ expression for the present elastoplastic model:

$$P = k_1 h^2 \tag{18}$$

where the frontal coefficient k_1 is expressed as the combination of the elastic measure E' and the plastic measure H:

$$k_1 = \frac{H}{(1 + \sqrt{2}/\tan\beta \sqrt{H/E'})^2} \left(\frac{g}{\gamma_1^2} \right) \tag{19}$$

where use has been made of the compatibility relation $\gamma_e \equiv \gamma_p$ between the elastic and plastic surface deformations, resulting in a "single" elastoplastic surface profile of impression with its geometric factor γ_1, namely $\gamma_1 = \gamma_e$ (= γ_p).

The Meyer hardness H_M is simply defined as the mean contact pressure given by the indentation contact load P divided by the projected area A_r of the residual impression after complete unloading:

$$H_M = \frac{P}{A_r} \tag{20}$$

Accordingly, hardness measurements quickly yield quantitative information on the characteristic mechanical properties such as the elastic, plastic and fracture resistance parameters. However, because of the multitude of relationships between the Meyer hardness and these mechanical parameters, some of which result from completely independent physical processes and mechanisms, the hardness measurement has been considered among the most maligned of physical measurements. For purely plastic indentation contact, due to none of elastic recovery of hardness impression during unloading, the contact area A_c at the peak load P_{max} remains intact, resulting in coincidence with the area of residual impression A_r. Accordingly, only for purely plastic indentation contact, comparing Eqs. (14) and (20), does H_M have a well-defined physical meaning of the true hardness H as a measure of plasticity, and then is linked to the yield strength Y through Eq. (15).

As noted in Section 2.3, in the conventional Vicker/Berkovich pyramid and also spherical indentation tests, the in-surface elastic recovery is negligibly small even if the in-depth elastic recovery is significant, implying that there is a good approximation of $A_c \cong A_r$ [53, 60, 61] and so affording the following alternative definition of the Meyer hardness:

$$H_M = \frac{P}{A_c} \tag{20'}$$

The experimental correlation between the values of Meyer hardness determined with both Eqs. (20) and (20') is plotted in Figure 2.4 for various types of materials ranging from brittle ceramics to ductile metals in indentation tests with pyramidal

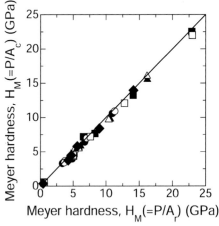

Figure 2.4 One-to-one correspondence of the Meyer hardness numbers determined from the projected contact area A_c (y-axis) and from the projected area A_r of residual impression (x-axis). The symbols indicate the experimental results obtained for various engineering materials ranging from brittle ceramics to ductile metals by the use of various types of trigonal and tetragonal pyramid indenters [61].

indenters (trigonal and tetragonal pyramids) having various values of inclined face angle β [61]. The excellent one-to-one correspondence illustrated in Figure 2.4 supports well the approximation $A_c \cong A_r$.

Substituting Eqs. (18) and (19) into Eq. (20') and utilizing the relations $A_c = gh_c^2$ and $h = \gamma_1 h_c$ enables the Meyer hardness of the present elastoplastic model to be given by the relation $H_M = k_1(\gamma_1^2/g)$ and then described in terms of E', H and the inclined face angle β of the cone/pyramid indenter as follows:

$$H_M \left(\equiv k_1 \frac{\gamma_1^2}{g} \right) = \frac{H}{(1 + \sqrt{2/\tan\beta}\sqrt{H/E'})^2} \tag{21}$$

As readily seen in Eq. (21), the Meyer hardness of elastoplastic material shows a somewhat complicated dependence on the elastic measure E', plastic measure H and the indenter's geometry through the inclined face angle β. The β-dependent nature of H_M means that the Meyer hardness is by no means a geometry-invariant characteristic material parameter. Some experimental results are shown in Figure 2.5; the normalized Meyer hardness H_M/H is plotted against the non-dimensional plastic parameter $(E'/H)\tan\beta_c$ for various materials from very brittle ceramics to ductile metals indented with trigonal or tetragonal pyramids having various inclined face angle β [61]. The β_c in Figure 2.5 is taken to be the angle of the "equivalent cone" which displaces the same volume of material indented by the respective pyramid indenters. Accordingly, this concept of equivalent cone results in $\beta_c = 19.7°$ for both the standard Berkovich pyramid with $\beta = 24.7°$ and the standard Vickers pyramid with $\beta = 22.0°$, as examples. The solid line

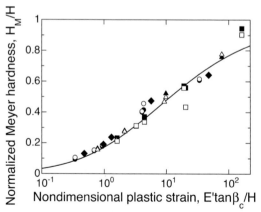

Figure 2.5 Normalized Meyer hardness plotted against the plastic strain for various engineering materials ranging from brittle ceramics to ductile metals by the use of various types of trigonal and tetragonal pyramid indenters [61]. The solid line indicates the theoretical prediction [Eq. (21)]. The inclined face angle of the equivalent cone indenter is represented by β_c.

in Figure 2.5 indicates the analytical prediction of Eq. (21), showing excellent coincidence between the experimental results and the prediction. Equation (21) suggests that the Meyer hardness H_M increases monotonically with the increases in E', H and also β. This is the reason why we implicitly expected a large E' value from its large H_M value observed when the test specimen is too small in dimension to measure the E' value by the conventional test techniques.

Equation (21) is reduced to the purely plastic relation of Eq. (14), that is, $H_M = H$, in the extreme of $H/E' \downarrow 0$, where the Meyer hardness signifies the true hardness as the measure of plasticity. In the other extreme of purely elastic contact ($H/E' \uparrow \infty$), Eq. (21) is recast into

$$H_M = \frac{E'}{2} \tan \beta \tag{22}$$

that is, the Meyer hardness of a purely elastic material represents the elastic measure of E' [5, 10].

As addressed in Section 2.3 [see Eq. (8)], the distorted surface profile of residual cone/pyramid impression gives rise to a non-quadratic $P - h$ unloading path of $P = \alpha(h - h_r)^m$ with $m \neq 2.0$. However, the geometrically quadratic nature of cone/pyramid indentation suggests that this non-quadratic unloading behavior can be expressed by adding a minor modification factor, $f(h_r/h_{max})$, to the quadratic expression, as follows [55, 63];

$$P = k_2(h - h_r)^2 f(h_r/h_{max}) \tag{23}$$

The details of the modification factor $f(h_r/h_{max})$ have been examined in a finite element analysis [63]. Since both the loading and unloading $P - h$ paths are basically quadratic, their $\sqrt{P} - h$ plots are supposed to be linear; two examples of $\sqrt{P} - h$ plots, for a float glass (soda-lime silicate glass) and a polycrystalline Si_3N_4 ceramic, are shown in Figure 2.6 (Vickers indentation), indicating that the unloading path of the Si_3N_4 ceramic is well approximated in a quadratic way, whereas a finite modification factor $f(h_r/h_{max})$ is required for the float glass. The distorted residual pyramid of Vickers impression (the pyramidal impression comprising convex side surfaces) gives rise to this non-quadratic nature to the unloading process of float glass [55]. The slope of the linear line connecting P_{max} at h_{max} and h_r (the dotted line in Figure 2.6) gives the unloading parameter $\sqrt{k_2}$, irrespective of quadratic unloading or not.

Noting that the unloading processes results mainly from the elastic recovery of impression and the unloading path coincides with the loading path in a purely elastic material [see Eqs. (12) and (13)], the unloading parameter k_2 can be well approximated by the following equation [55, 63]:

$$k_2 = \frac{E' \tan(\beta - \beta_r)}{2} \frac{g}{\gamma_e^2} \tag{24}$$

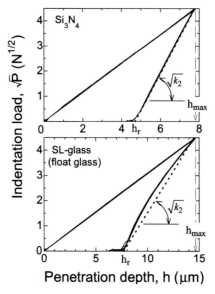

Figure 2.6 $\sqrt{P}$ vs. h linear plots for an Si_3N_4 ceramic and a float glass for Vickers indentation. The unloading parameter $\sqrt{k_2}$ is defined as the slope of the linear unloading lines (the dotted lines).

where β_r is the inclined face angle of residual impression. The indenter's geometric factor g is $\pi \cot^2(\beta - \beta_r)$, $3\sqrt{3}\cot^2(\beta - \beta_r)$ and $4\cot^2(\beta - \beta_r)$ for a cone, trigonal pyramid and tetragonal pyramid indenter, respectively. Equation (24) was derived theoretically from the fact that the elastic contact process of a rigid cone/pyramid indenter having a face angle β and a residual impression with the face angle β_r is equivalent to the elastic process of a rigid cone/pyramid indenter having an effective face angle β_{eff} ($\equiv \beta - \beta_r$) pressed into contact with a flat elastic half-space. Rewriting Eq. (24) with respect to the elastic modulus E' and utilizing Love's relation $\gamma_e = \pi/2$, we have the following equation, which is efficient practically for determining experimentally the elastic modulus E' from the unloading parameter k_2 [55, 63]:

$$E' = \frac{\pi^2}{2g\tan(\beta - \beta_r)}k_2 \tag{25}$$

The experimental determination of β_r can be done, without directly observing the residual impression, by the use of the relative residual depth ξ_r ($= h_r/h_{max}$) that is an observable parameter in $P - h$ or $\sqrt{P} - h$ hysteresis plots (see Figures 2.3 and 2.6), using the approximation $\tan \beta_r \cong \xi_r \tan \beta$. The details have been reported in the literature [55, 63]. The elastic modulus E' thus estimated in this way using Eq. (25) is plotted in Figure 2.7 against the value determined using the conventional

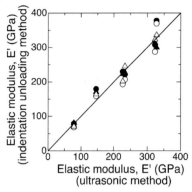

Figure 2.7 One-to-one correspondence of the elastic moduli determined from the indentation unloading data (y-axis) and from the conventional ultrasonic method (x-axis) for various engineering materials ranging from brittle ceramics to ductile metals [61].

ultrasonic method, indicating an excellent one-to-one correlation for various types of brittle and ductile materials contacted with various types of pyramidal indenters [61].

The depth of the residual impression h_r (see Figures 2.3 and 2.6) may be an efficient measure of the plasticity of material indented; a purely elastic indentation results in $h_r = 0$ and a purely plastic indentation yields $h_r = h_{max}$ (no elastic recovery). It has been well recognized that the relative expression of h_r normalized with the maximum penetration depth h_{max}, that is, $\xi_r (= h_r/h_{max})$, is independent of the load and penetration depth of pyramidal indentation, suggesting a characteristic material parameter representing the degree of plasticity; ranging from 0.0 for purely elastic to 1.0 for purely plastic materials [42, 51, 53, 61]. The experimental relationship between $\xi_r (= h_r/h_{max})$ and the dimensionless plastic strain $E'\tan\beta_c/H$ is illustrated in Figure 2.8 (closed circles) for various brittle to ductile materials indented by pyramidal indenters having various inclined face angle β. As can be clearly seen, all of the materials pressed into contact with various types of pyramidal indenter are superimposed into a single master line as a function of the plasticity $E'\tan\beta/H$, implying that ξ_r can also be utilized as a characteristic measure of elastoplastic contact deformations.

Combining the quadratic $P - h$ loading relation $P = k_1 h^2$ and the quadratic unloading relation $P = k_2(h - h_r)^2$ along with the mechanical compatibility at h_{max}, namely $k_1 h_{max}^2 = k_2(h_{max} - h_r)^2$, and taking account of the expressions in Eqs. (19) and (24), the relative residual depth ξ_r is given by the following equation in terms of the plasticity $E'\tan\beta/H$:

$$\xi_r = \frac{1 + \sqrt{\dfrac{2H}{E'\tan\beta}}\left(1 - \sqrt{\dfrac{\tan\beta_{eff}}{\tan\beta}\dfrac{\gamma_e}{\gamma_1}}\right)}{1 + \sqrt{\dfrac{2H}{E'\tan\beta}}} \qquad (26)$$

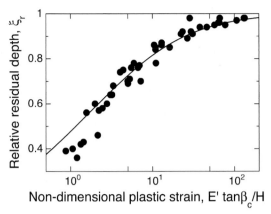

Figure 2.8 Relative residual depth as a function of plastic strain for various engineering materials ranging from brittle ceramics to ductile metals by the use of various types of trigonal and tetragonal pyramid indenters. The solid line indicates the theoretical prediction [Eq. (26)]. The inclined face angle of the equivalent cone indenter is represented by β_c.

where we have the relation $\tan \beta_{\text{eff}}/\tan \beta = (1 - \xi_r)/[1 + \xi_r (\tan \beta)^2]$ through the approximation $\tan \beta_r \cong \xi_r \tan \beta$, and also γ_e/γ_i is described in terms of ξ_r by Eq. (32) (see Section 2.4.4). Since the relative residual depth ξ_r is included in both sides of Eq. (26), the numerical solution of ξ_r as a function of the plastic strain $E' \tan \beta/H$ can be obtained in an iterative manner and is indicated in Figure 2.8 by the solid line. There is excellent quantitative agreement between the experimental observations (closed circles) and the theoretical prediction of the present simple elastoplastic model.

2.4.4
Indentation Contact Area A_c; Oliver–Pharr/Field–Swain Approximation

As mentioned in the preceding sections, the elastic recovery of the hardness impression dominates its unloading $P - h$ path. As actually realized in Eq. (24), the unloading parameter k_2 is directly linked to the elastic modulus E'. The technique for determining E' from the unloading stiffness $S(= dP/dh)$ (the slope of unloading $P - h$ curve at P_{max}) relies on the fact that the displacement recovered during unloading is largely elastic, in which case the purely elastic contact mechanics is supposed to be satisfactorily used. The quadratic unloading path $P = k_2(h - h_r)^2$ for *conical/pyramidal indentation* combined with the elastic expression of Eq. (24) results in the following relationship between the unloading stiffness S and the indentation contact area A_c at P_{max}:

$$S = \frac{2}{\sqrt{\pi}} \sqrt{A_c} E' \qquad (27)$$

Pharr, Oliver and Brotzen demonstrated that for an indenter of solid of revolution (cone, sphere, paraboloid of revolution and so forth), Eq. (27) is an indenter's geometry-independent relation among S, A_c and the elastic modulus E', if the indenter is rigid enough (if the deformation of the indenter is negligibly small) [40]. Accordingly, we can estimate the elastic modulus E' of the material indented using Eq. (27) and also the Meyer hardness H_M through Eq. (20'), *when we know the contact area A_c at h_{max}.*

Among the several empirical techniques and approximations/assumptions for estimating A_c is that the contact periphery sinks-in in a manner that can be given by rigid punches having simple geometries pressed into contact with a flat elastic half-space. This elastic assumption includes a critical limit, as it does not account for the pile-up of material at the contact periphery that occurs in ductile materials. The unloading processes are shown schematically in Figure 2.9 for hardness impression with a sinking-in profile, where a cone indentation is demonstrated as an example of indenters of solid of revolution (cone/pyramid, sphere, paraboloid of revolution and so forth). Figure 2.9a shows the unloading process of an elasto-plastic indentation; the contact depth h_c and the sink-in depth h_s yield the total penetration depth h_{max} at the maximum load P_{max}, leaving the residual depth h_r after the elastically recovered penetration h_e. There are the following relations between these penetration depths: $h_{max} = h_s + h_c = h_r + h_e$ (see Figure 2.9a).

Consider a purely elastic indentation as shown in Figure 2.9b, where *the total penetration depth at P_{max} is supposed to equal the recovered elastic depth h_e of the elasto-plastic indentation given in Figure 2.9a. The sink-in depth and the contact depth*

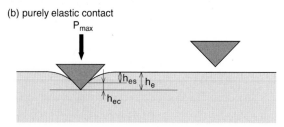

(a) elastoplastic contact

(b) purely elastic contact

Figure 2.9 Contact surface deformations for (a) elastoplastic and (b) purely elastic indentations at the maximum load and after complete unload.

of this elastic indentation are expressed by h_{es} and h_{ec}, respectively, through the relation $h_e = h_{es} + h_{ec}$. The following elastic assumption was made by Oliver and Pharr for pyramidal indentation [41] and by Field and Swain for spherical indentation [43]:

$$h_s = h_{es} \tag{28}$$

In other words, *the sink-in depth of elastoplastic impression is supposed to be equal to that of the elastic impression when the depth of elastic recovery in elastoplastic indentation is the same as that of the total penetration depth in purely elastic indentation.* Combining Eq. (28) with the preceding relations ($h_{max} = h_s + h_c = h_r + h_e$; $h_e = h_{es} + h_{ec}$) for the respective penetration depths in elastoplastic and purely elastic indentations, the elastoplastic contact depth h_c can be given in terms of the experimentally observable depths h_r and h_{max}, as follows:

$$h_c = \left(1 - \frac{1}{\gamma_e}\right)h_r + \frac{1}{\gamma_e}h_{max} \tag{29}$$

where γ_e represents the surface profile of elastic impression, being defined by $\gamma_e = h_e/h_{ec}$ in Figure 2.9b (the ratio of the total penetration depth to the contact depth in elastic contact). The γ_e value is 2.0 for spherical contact (Hertzian contact), leading to the Field–Swain approximation [43], and $\pi/2$ for conical/pyramidal indentation, leading to the Oliver–Pharr approximation [41].

Under the assumption of Eq. (28) and the relations $h_{max} = h_s + h_c$ and Eq. (29), the sink-in depth h_s of elastoplastic indentation can be described by

$$h_s = \left(1 - \frac{1}{\gamma_e}\right)(h_{max} - h_r) \tag{30}$$

in terms of the experimentally observable depths h_r and h_{max}. The $P - h$ unloading expressions in Eqs. (8) and (30) result in the following Oliver–Pharr relation [41]:

$$h_s = m\left(1 - \frac{1}{\gamma_e}\right)\frac{P_{max}}{S} \tag{31}$$

In the original Oliver–Pharr expression, the frontal factor was expressed with ε as $\varepsilon = m(1 - 1/\gamma_e)$. Accordingly, it results in $\varepsilon = 0.726$ (that is, $m = 2.0$ and $\gamma_e = \pi/2$) for cone/pyramid indentation and $\varepsilon = 0.750$ (that is, $m = 3/2$ and $\gamma_e = 2.0$) for spherical indentation.

Equation (29) (Oliver–Pharr/Field–Swain approximation) can be recast into the following alternative expression [53]:

$$\frac{1}{\gamma_1} = \frac{1}{\gamma_e} + \left(1 - \frac{1}{\gamma_e}\right)\xi_r \tag{32}$$

using the elastoplastic impression profile γ_1 defined by $\gamma_1 = h_{max}/h_c$, indicating that the γ_1 value approaches its elastic value of γ_e when ξ_r meets the elastic extreme of 0.0, whereas the γ_1 value decreases monotonically with increase in the relative residual depth ξ_r, reducing to 1.0 in the plastic extreme of $\xi_r = 1.0$, *suggesting no piling-up profile of impression even in the perfectly plastic extreme.*

Equation (32) (Oliver–Pharr/Field–Swain approximation) for *pyramidal indentation* is plotted in Figure 2.10 as the solid line along with the experimental observations (filled symbols) [53]. As can be clearly seen, the theoretical prediction of Eq. (32) (solid line) always overestimates the γ_1 value (and underestimates h_c and hence A_c), due to the critical assumption (the assumption of always a sinking-in profile of impression) included in Eqs. (28), (29) and (32). The empirical line best fitted to the observations is indicated by the broken line in Figure 2.10; its empirical equation (cone/pyramid indentation) is [53, 61]

$$\gamma_1 = \frac{\pi}{2}(1 - 0.43\sqrt{\xi_r}) \tag{32'}$$

The discrepancy observed between the solid line (Oliver–Pharr/Field–Swain approximation) and the broken line (empirically best-fitted line) signifies the limit of the Oliver–Pharr/Field–Swain approximation. The indentation contact area A_c that is required for calculating the Meyer hardness [Eq. (20')] and the elastic modulus [Eq. (27)] can be obtained from the observable penetration depth h_{max} combined with the γ_1 value [Eq. (32) or (32')] using the following equations:

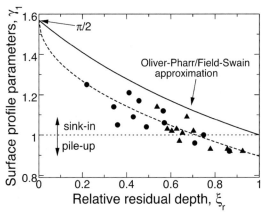

Figure 2.10 Surface profile parameter γ_1 ($= h_{max}/h_c$) plotted against the relative residual depth ξ_r ($= h_r/h_{max}$) for various engineering materials ranging from brittle ceramics to ductile metals by the use of various types of trigonal and tetragonal pyramid indenters. The solid line indicates the Oliver–Pharr/ Field–Swain approximation [Eq. (32) with $\gamma_e = \pi/2$]. The broken line is best fitted to the observations (symbols), its empirical equation being given as Eq. (32'). The impression profile is sinking-in for $\gamma_1 > 1.0$ and piling-up for $\gamma_1 < 1.0$ [53].

$$A_c = \frac{g}{\gamma_1^2}(h_{max})^2 \tag{33}$$

for cone/pyramid indentation (the g value is 24.5 for Vickers and Berkovich indentation) and

$$A_c = \frac{2\pi R}{\gamma_1} h_{max} \tag{33'}$$

for spherical indentation (the contact radius a is assumed to be small enough compared with the radius R of the sphere). In spherical indentation, Eq. (32') (the empirical equation applicable only to pyramidal indentation) must be replaced with the following empirical expression that was derived from a simple modification of Eq. (32) by taking account of the pile-up of impression for ductile materials [64]:

$$\frac{1}{\gamma_1} = \frac{1}{\gamma_e} + \left(1.3 - \frac{1}{\gamma_e}\right)\xi_r \tag{32''}$$

with $\gamma_e = 2.0$. Accordingly, using Eq. (33) or (33'), we can easily estimate the indentation contact area A_c from the empirically observed elastoplastic characteristic parameter of the relative residual depth $\xi_r(= h_r/h_{max})$ combined with the approximation Eq. (32) or the empirical equations Eq. (32') or (32'').

2.4.5
Energy Principle of Indentation Contact [42]

The hysteresis loop energy U_r enclosed with the loading and unloading paths (see Figure 2.3) is the indentation work consumed in the indentation loading–unloading cycle. This irreversible energy is consumed to create the elastoplastic hardness impression. Accordingly, this hysteresis loop energy includes important information on the elastoplastic characteristics of the material indented.

The external work (total indentation energy, U_t) applied to the material indented to the maximum penetration depth h_{max} is given by the integral of indentation load P with respect to the penetration depth h along the loading path, that is, $U_t = \int_0^{h_{max}} Pdh$ and then Eq. (18) results in

$$U_t = \frac{1}{3}k_1(h_{max})^3 \tag{34}$$

for the cone/pyramid indentation. The elastic energy U_e released during unloading that is associated with the elastic recovery of penetration depth from h_{max} to h_r is expressed by the integral of indentation load P with respect to the penetration depth h along the unloading path, that is, $U_e = \int_{h_r}^{h_{max}} Pdh$, resulting in

$$U_e = \frac{1}{3} k_2 (h_{max})^3 (1 - \xi_r)^3 \tag{35}$$

where use has been made of the quadratic expression for the unloading $P - h$ relation $P = k_2(h - h_r)^2$. Accordingly, the hysteresis loop energy U_r consumed in creating the residual hardness impression is given by

$$U_r \left(\equiv U_t - U_e \right) = \frac{1}{3} k_1 (h_{max})^3 \xi_r \tag{36}$$

In the derivation of Eq. (36), we used the mechanical compatibility relation $\sqrt{k_1} h_{max} = \sqrt{k_2} (h_{max} - h_r)$ (or its alternative expression $\xi_r = 1 - \sqrt{k_1/k_2}$).

Among the energy-derived indentation parameters is the non-dimensional measure of ductility (indentation ductility index) D, defined as the ratio of the irreversible hysteresis energy U_r to the total energy applied to the maximum depth of penetration U_t:

$$D = \frac{U_r}{U_t} \tag{37}$$

D must be 0.0 for a purely elastic body with complete elastic recovery during unloading ($U_r = 0.0$) and 1.0 for a purely ductile material ($U_r \equiv U_t$). Substituting Eqs. (34) and (36) into Eq. (37) yields the simple relation $D = \xi_r$, indicating that the ductility index D is quantitatively described by the elastic modulus E', true hardness H and the geometry of the indenter (the inclined face angle β for cone/ pyramid indentation, for example) through a function of the plastic strain of (E' $\tan \beta)/H$ [see Eq. (26) and Figure 2.8]. One of the practical applications of the ductility index D was made through the combination with the energy-derived fracture toughness $(K_{Ic})^2/E'$ for examining the machinability and the machining-induced contact damages of engineering ceramics [65].

Another important elastoplastic parameter that is derived from the energy-based indentation characteristics is the work of indentation, W_I, defined as the indentation work required to create a unit volume of residual impression:

$$W_I = \frac{U_r}{V_r} \tag{38}$$

where V_r is the volume of residual impression, expressed by

$$V_r = \frac{1}{3} \frac{g}{\gamma_1^2} (h_{max})^3 \xi_r \tag{39}$$

under the assumptions that the side surfaces and edges of pyramid impression are straight and the projected contact area A_c at P_{max} coincides with A_r of the residual impression (no in-surface elastic recovery; see the preceding consider-ations in Sections 2.3 and 2.4.3). On combining Eqs. (36), (38) and (39) and com-paring the result with Eq. (21), it is worthy of note that the work of indentation coincides with the Meyer hardness:

$$W_I \equiv H_M \tag{40}$$

Accordingly, it can be concluded from the present energy-based considerations that *the Meyer hardness is the elastoplastic work for creating a unit volume of hardness impression.*

Equation (21) implies that the Meyer hardness H_M is reduced to the true hard-ness H in the purely plastic/ductile extreme of $(E' \tan \beta)/H \uparrow \infty$ or $D(\equiv \xi_r) \uparrow 1.0$. Accordingly, in the extreme of ductile flow, the work of indentation affords the true hardness and then the yield strength Y through the relation $Y = H/C$ [see Eq. (15)]. In other words, *the true hardness H is the indentation work to create a unit volume of purely plastic impression.*

2.5
Contact Mechanics of the Coating–Substrate System

One of the most appropriate applications of micro/nanoindentation technique is the mechanical characterization of coating films such as those used in MEMS, surface modification for optoelectronic devices, microelectromagnetic devices and so forth. Among numerous advantages in indentation mechanical testing, the technique does not require any specific preparation and geometry of the test speci-men, in addition to the essential capability of determining the characteristic mechanical properties of thin films and materials in volumes with extremely small dimensions of several tens of nanometers or less. On applying micro/nanoinden-tation techniques to characterizing thin films coated on a substrate, the most criti-cal difficulty is the substrate effect on the mechanical properties of the coating. If the substrate effect is not properly taken into account, significant errors and ambi-guity will be included in the observed indentation $P - h$ hysteresis curve, resulting in yielding wrong information on the film-only properties. Several models have been proposed and theoretical/experimental studies conducted to account for film–substrate effects on elastic modulus and hardness measurements in indenta-tion tests [35, 66–96]. However, rigorous test techniques and procedures have not yet been established for determining the film-only characteristic mechanical prop-erties. Furthermore, the recent advances in micro/nanotechnologies require thin films having the thicknesses of about 100 nm or less, implying that there is a need to develop novel indentation test methods and analyses that can properly eliminate the substrate effect from the observed data, as the penetration depth h required in nanoindentation test techniques has the same order of the thickness as coatings.

2.5.1
Elastic Indentation Contact Mechanics

A number of analytical studies have been reported on axisymmetric indenters pressed into contact with elastic coating–substrate systems. The contact mechanics problem for an elastic layer resting on a rigid foundation in a frictionless manner was solved for a flat-ended cylinder by Dhaliwal [66], and Dhaliwal and Rau [67] subsequently extended this analysis to axisymmetric indenters with an arbitrary profile pressed into an elastic coating perfectly bonded to or frictionlessly overlying an elastic semi-infinite substrate. The coordinate systems and the contact parameters are depicted in Figure 2.11a–c for a flat-ended cylinder (radius a), sphere (radius R) and cone/pyramid (inclined face angle β) indentations, respectively. These Boussinesq problems for elastic contact have been reduced to a single Fredholm integral equation of the second kind [66, 67, 70, 87]:

$$H(r/a) - \frac{1}{\pi}\int_0^1 K(y; r/a, t_f/a)H(y)dy = F(r/a; a/a_H) \tag{41}$$

where $H(r/a)$, $F(r/a; a/a_H)$ and $K(y; r/a, t_f/a)$ are, respectively, the non-dimensional contact pressure distribution, dimensionless geometry of axisymmetric indenter and the kernel of the integral equation including information on the characteristic material parameters both of the coating–substrate system and the contact boundary conditions. As indicated in Figure 2.11, the thickness of the film is represented by t_f. These Boussinesq problems are described in cylindrical coordinates (r,θ,z) such that the z-coordinate coincides with the axis of the indenter, r is perpendicu-

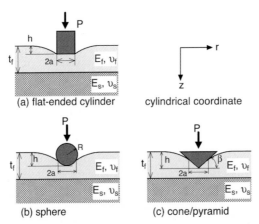

Figure 2.11 Contact parameters, coordinate system and surface deformations of coating–substrate composites contacted with (a) flat-ended cylinder, (b) sphere and (c) cone/pyramid indenters.

lar to z and θ is the azimuthal angle. The origin of the coordinates is located at the first contact point of the indenter on the surface of the flat layer (the center of contact circle for the flat-ended cylinder [Figure 2.11(a)]. The contact radius of the coating–substrate system is expressed by a. Due to a significant substrate effect, the contact radius a is, in general, totally different from the contact radius a_{H} for the *homogeneous* semi-infinite half-space [70]. For example, a spherical indentation on a soft coating–hard substrate system results in an enhancement of pile-up even in elastic contact; the material of the coating elastically deformed beneath the indenter is constrained by the rigid substrate, enhancing the elastic deformations upwards to the regions near the contact periphery, subsequently increasing the amount of elastic pile-up. In contrast, for a coating overlying a compliant substrate, a sink-in contact profile is more enhanced than that for the *homogeneous* semi-infinite half-space. Accordingly, a is larger than a_{H} ($a > a_{\mathrm{H}}$) in the former and smaller ($a < a_{\mathrm{H}}$) in the latter for a given penetration depth.

The Fredholm integral equation of the second kind [Eq. (41)] was solved numerically by Yu et al. in the form of a Chebyshev series in a precise manner [70]. However, this analytical/numerical approach requires an extremely tedious and complicated mathematical algorithm, reducing the importance of it highly reliable numerical predictions for elastic contact deformations of coating–substrate systems.

To circumvent the mathematical and numerical complexities included in Eq. (41) from the viewpoint of practical examinations for the elastic contact deformations of coating–substrate systems, use has been made of Boussinesq's Green function $G(r, z; E, v)$, that is, the elastic solution of the contact surface deformation $w(r, z)$ for a point loading P on a *homogeneous* semi-infinite half-space (elastic modulus E and Poisson's ratio v) [3]:

$$G(r, z; E, v)[\equiv w(r, z)] = \frac{P}{2\pi}\frac{1+v}{E}\left[\frac{r^2}{\sqrt{(r^2+z^2)^3}} + \frac{3-2v}{\sqrt{r^2+z^2}}\right] \tag{42}$$

Applying the Green function $G(r, z; E, v)$ for a concentrated load P acting on the surface of a semi-infinite body to a coating–substrate layered composite, we can find the surface displacements and then the penetration depth h induced by a distributed contact stress $p(r; a)$ beneath a solid indenter using the superposition procedure, as follows [89, 93]:

$$h = \int_0^a p(r, a)\left[\int_0^{t_{\mathrm{f}}} G(r, z; E_{\mathrm{f}}, v_{\mathrm{f}})dz + \int_{t_{\mathrm{f}}}^\infty G(r, z; E_{\mathrm{s}}, v_{\mathrm{s}})dz\right]2\pi r dr \tag{43}$$

where the subscripts f and s indicate the film and the substrate, respectively. Integration of Eq. (43) gives the following general expression between h and P for axisymmetric indentation geometries:

$$h = \frac{A}{E'_{\mathrm{eff}}/E'_{\mathrm{f}}}\frac{P}{aE'_{\mathrm{f}}} \tag{44}$$

where $A = 0.5$, 0.75 and 1.0 for flat-ended cylinder, sphere and cone indentations, respectively [93]. The normalized elastic modulus $E'_{\text{eff}}/E'_{\text{f}}$ (elastic modulus of coating–substrate composite system E'_{eff} normalized by the film modulus E'_{f}) in Eq. (44) can then be given by [96]

$$\frac{E'_{\text{eff}}}{E'_{\text{f}}} = \frac{1}{1 + \dfrac{1}{2I_0}\dfrac{1}{1-v}[(3-2v)I_1(t_{\text{f}}/a) - I_2(t_{\text{f}}/a)]\left(\dfrac{E'_{\text{f}}}{E'_{\text{s}}}-1\right)} \tag{45}$$

with the functions $I_1(t_{\text{f}}/a)$ and $I_2(t_{\text{f}}/a)$ defined by

$$I_1(t_{\text{f}}/a) = \int_0^1 p_N(x)\frac{x}{\sqrt{x^2+(t_{\text{f}}/a)^2}}\,dx \tag{46}$$

and

$$I_2(t_{\text{f}}/a) = \int_0^1 p_N(x)\frac{x^3}{\left[\sqrt{x^2+(t_{\text{f}}/a)^2}\right]^3}\,dx \tag{47}$$

where x represents the normalized radius r/a and $p_N(x)$ is the normalized contact pressure distribution defined by $p_N(x) = p(r,a)/p_{\text{m}}$ in terms of the mean contact pressure $p_{\text{m}} = P/(\pi a^2)$. For the extreme of thin coating of $t_{\text{f}}/a = 0.0$, both of the functions $I_1(t_{\text{f}}/a)$ and $I_2(t_{\text{f}}/a)$ coincide with each other, giving the value of $I_0 = \left[= I_1(0) = I_2(0) = \int_0^1 p_N(x)dx\right]$. In Eq. (45), for simplicity, the Poisson's ratios v_{f} and v_{s} are supposed to be similar in magnitude, being assumed to be equal to a single value of v. Equation (45) implies that we can *analytically* describe the composite modulus E'_{eff} as a function of the normalized film thickness t_{f}/a, once we know the modulus ratio $E'_{\text{f}}/E'_{\text{s}}$ or determine the film modulus E'_{f} once we have measured the composite modulus E'_{eff} in an indentation test provided that the substrate modulus E'_{s} is known.

The effective modulus E'_{eff} of a coating–substrate composite can be expressed in a phenomenological manner using the following stiffness weight function $W(t_{\text{f}}/a)$ [72, 73]:

$$E'_{\text{eff}} = E'_{\text{s}} + (E'_{\text{f}} - E'_{\text{s}})W(t_{\text{f}}/a) \tag{48}$$

The stiffness weight function $W(t_{\text{f}}/a)$ is related to the normalized elastic modulus $E'_{\text{eff}}/E'_{\text{f}}$ [Eq. (45)] of the coating–substrate composite in terms of the modulus ratio $E'_{\text{f}}/E'_{\text{s}}$ as follows:

$$W(t_{\text{f}}/a) = \frac{1-(E'_{\text{f}}/E'_{\text{s}})(E'_{\text{eff}}/E'_{\text{f}})}{1-(E'_{\text{f}}/E'_{\text{s}})} \tag{48'}$$

The weight function $W(t_{\text{f}}/a)$ increases monotonically from 0 to 1.0 when the normalized film thickness t_{f}/a changes from 0.0 (substrate only) to infinity (film only).

In spherical indentation as an example, the stiffness weight function $W(t_f/a)$ thus defined is plotted against the normalized film thickness t_f/a in Figure 2.12 for the numerical solutions of the Fredholm integral equation [Eq. (41), symbols] and for the analytical solutions [Eq. (45) combined with Eq. (48'), solid lines] with various values of the modulus ratio E_f'/E_s'. In calculating analytically or numerically the functions $I_1(t_f/a)$ and $I_2(t_f/a)$ [Eqs. (46) and (47) and also the I_0 value], use has been made of the homogeneous approximation for the normalized contact pressure distribution $P_N(x)$ for spherical indentation (that is, the contact pressure distribution induced in the coating–substrate layers is assumed to be the same as the distribution generated in a homogeneous semi-infinite half-space). This approximation results in $P_N(x) = (3/2)\sqrt{1-x^2}$ for spherical indentation. The details of this homogeneous approximation and its limit have been intensively examined in the literature [93, 96].

As can be clearly seen in Figure 2.12, the analytical results [Eq. (45)] reproduce well the rigorous solutions of the Fredholm integral equation [Eq. (41)]. This suggests that we can estimate analytically the elastic modulus of the coating film E_f' when we know the composite modulus E_{eff}' determined in a micro/nanoindentation test in terms of the dimensionless film thickness t_f/a. However, in this experimental procedure for estimating E_f', there exists a crucial problem in estimating the contact radius a of the coating–substrate composite system. As mentioned above, the substrate significantly affects the contact radius a. In cases of an axisymmetric indenter pressed into contact with a homogeneous elastic half-space, the contact radius a_H is given analytically by the following simple equation in terms of the given penetration depth h:

$$a_H = Bh^{n-1} \tag{49}$$

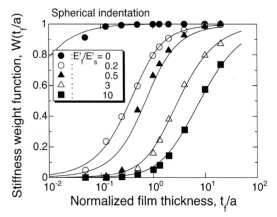

Figure 2.12 on axes: Spherical indentation. Y-axis: Stiffness weight function, $W(t_f/a)$. X-axis: Normalized film thickness, t_f/a. Legend: $\bullet$: $E_f'/E_s' = 0$; $\circ$: 0.2; $\blacktriangle$: 0.5; $\triangle$: 3; $\blacksquare$: 10.

Figure 2.12 Stiffness weight function (spherical indentation) as a function of the normalized film thickness for various values of the modulus ratio E_f'/E_s'. The symbols are the numerical results of Eq. (41) and the solid lines are the analytical predictions of Eq. (48').

where the exponent n is 1.0, 1.5 and 2.0 and the frontal coefficient B is a (radius of flat-ended cylinder), $\sqrt{R}$ and $2\cot\beta/\pi$ for a flat-ended cylinder, sphere and cone indentation, respectively [93].

The Fredholm integral equation [Eq. (41)] not only yields E'_{eff}/E'_f (see the symbols in Figure 2.12 for spherical indentation as an example) but also provides the normalized contact radius a/a_H as a function of t_f/a through solving the mixed boundary Fredholm integral equation by a long iterative procedure. To circumvent such an impractical situation for the mechanical characterization of a coating–substrate system, much simpler or analytical approaches have long been desired to estimate the contact radius a of layered composites. However, no analytically closed expressions for the normalized contact radius a/a_H of layered composites have been reported. Hsueh and Miranda examined an "empirical" equation for a/a_H by best fitting this function to the results of finite element analysis (FEA) for spherical indentation [90]. Sakai et al. extended their equation to include other axisymmetric indentation contact problems [93, 96]:

$$\frac{a}{a_H} = \left(\frac{E'_{eff}}{E'_f}\right)^{(n-1)/n}\left[1 - \frac{1-(E_f/E_s)^{(n-1)/n} + \chi(t_f/a; E_f/E_s)}{1+\Sigma_j C_j(t_f/a)^j}\right] \qquad (50)$$

where $\chi(t_f/a; E_f/E_s)$ and C_j are the fitting function and parameters, respectively, the details of which have been given in the literature [90]. The exponent n in Eq. (50) is 1.0, 1.5 and 2.0 for flat-ended cylinder, sphere and cone indentations, respectively.

Figure 2.13 illustrates the relationship between a/a_H and t_f/a, in which the rigorous solutions of the Fredholm integral equation [Eq. (41); symbols] and the

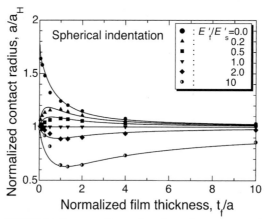

Figure 2.13 Contact radius a of coating–substrate composite normalized with the radius a_H for a homogeneous semi-infinite half-space plotted against the normalized film thickness for various values of the modulus ratio E'_f/E'_s in spherical indentation. The symbols are the numerical results of Eq. (41) and the solid lines are the predictions of Eq. (50).

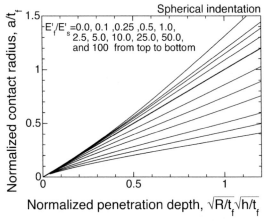

Figure 2.14 Relationship between the contact radius a and the penetration h in their normalized expressions for spherical indentation [see Eq. (51)].

empirical expression [Eq. (50); solid lines] are compared. Except for the coating overlying a very compliant substrate with $E_f'/E_s' \geq 10$ in the regions of extremely thin film or very deep penetration of $t_f/a \leq 0.5$, the empirical expression Eq. (50) represents well the rigorous solution of the Fredholm integral equation. As can be clearly seen in Figure 2.13, when the substrate is stiffer than the film ($E_f'/E_s' < 1.0$), the contact radius a of the layered composite is always larger than that of homogeneous half-space (that is, $a/a_H > 1.0$), whereas the ratio a/a_H is always smaller than 1.0 for a film coated on a compliant substrate ($E_f'/E_s' > 1.0$).

Combining Eq. (49) with Eq. (50) enables the contact radius a of layered composite to be obtained in terms of the penetration depth h that is an observable parameter in indentation tests. The a versus h relationship thus analytically derived is plotted in Figure 2.14 for *spherical indentation*, as an example, according to the dimensionless equation

$$\frac{a}{t_f} = \left(\frac{a}{a_H}\right)\sqrt{\frac{R}{t_f}} \times \sqrt{\frac{h}{t_f}} \tag{51}$$

Equation (51) and its numerical results shown in Figure 2.14 are essential from a practical point of view, as we can estimate the unobservable contact radius a from the experimentally observable penetration depth h and then using the dimensionless film thickness t_f/a, the composite modulus E_{eff}' as a function of h and so the indentation load P versus penetration depth h relation for the coating–substrate composite can be predicted analytically.

Figure 2.15 illustrates the elastic $P - h$ loading curves of methylsilsesquioxane films having various thicknesses t_f ranging from 0.5 to 20 μm coated on a soda-lime silicate glass plate (float glass plate) contacted with spherical indenters (with tip radius R of 4.5 and 50.0 μm) [93]. The solid lines in Figure 2.15 indicate the ana-

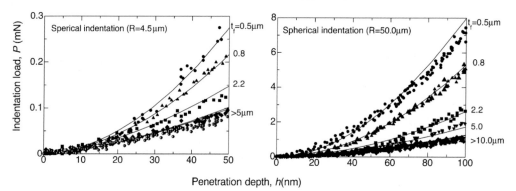

Figure 2.15 *P – h* loading curves of spherical indentations (tip radius of 4 and 50 μm) on an organic–inorganic hybrid film (methylsilsesquioxane film) coated on a soda-lime glass having various values of the film thickness. The solid lines are the analytical predictions [$P = (4/3)\sqrt{R}E'_{\text{eff}}h^{3/2}$ combined with Eqs. (45), (50) and (51)].

lytical prediction, showing excellent agreement with the results observed in the spherical nanoindentation tests. This excellent coincidence between observation and the analytical prediction implies that one can estimate the elastic modulus of coating film E'_f from the observed $P – h$ loading curve through inversely solving the analytical problem demonstrated in this section for estimating the $P – h$ loading curve of a coating–substrate system.

The detailed experimental procedures for estimating the elastic modulus of a coating film E'_f from the observed composite modulus E'_{eff} in a micro/nanoindentation test are shown schematically in Figure 2.16 for spherical indentation, as an example. Figure 2.16a is an observed elastic $P – h$ loading curve, in which a specific set of P and h are chosen to calculate the corresponding specific value of E'_{eff} at a specific penetration h. Figure 2.16b, which is a schematic of Figure 2.14, enables one to determine the specific contact radius a induced by the specific penetration h by *assuming* an appropriate value of the film modulus E'_f and so *assuming* an appropriate ratio E'_f/E'_s (the modulus of the substrate is supposed to be known) in this *first run*. The specific contact radius a thus estimated allows the determination of the stiffness weight function $W(t_f/a)$ at a specific value of a/t_f in Figure 2.16c (see Figure 2.12 for the same E'_f/E'_s ratio assumed). Substituting this specific $W(t_f/a)$ value and the effective modulus E'_{eff} that is determined from the observed $P – h$ curve into the equation $E'_{\text{eff}} = E'_s + (E'_f - E'_s)W(t_f/a)$, the film modulus E'_f can be determined, because the modulus of the substrate E'_s is known *a priori*. If the E'_f value thus determined shows a finite discrepancy from the value assumed in this first run, the E'_f value determined in the first run is utilized to determine newly the E'_f value in the second run in an iterative manner, until the estimated

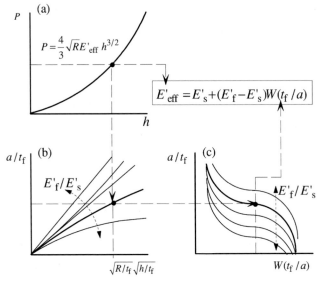

Figure 2.16 Iterative procedure for estimating the film modulus E'_f in spherical indentation test.

E'_f value converges to a specific value. In general, three or four cycles of iteration yield a reliable value of the film modulus E'_f.

Furthermore, through conducting the separate indentation tests using two spherical indenters having different tip radii R_1 and R_2, we can determine simultaneously the film modulus E'_f and its thickness t_f, if the elastic modulus of the substrate E'_s is known, and also make a simultaneous determination of E'_f and E'_s, if the film thickness is known [93].

2.5.2
Elastoplastic Indentation Contact Mechanics

The contact stresses induced by a conical/pyramidal indenter are highly concentrated beneath the tip of the indenter, inducing plastic flows to avert this stress singularity, even when the penetration is infinitesimally small. Plastic flows in hardness impression yield an additional difficulty in the indentation contact mechanics of a coating–substrate system. As is readily seen in Figure 2.14 for elastic indentation, the substrate-affected contact radius a is estimated analytically from the observable penetration depth h, enabling one to determine the elastic properties of layered composites in an indentation test (see Figure 2.16). For a soft and ductile film coated on a hard and brittle substrate as an example, however, the indentation-induced plastic flow greatly enhances the pile-up of impression, resulting in a significant modification of the a value that has been estimated for a purely elastic contact. The enhancement of pile-up of the film results from the material displaced by the indenter that is spatially constrained against the hard substrate to flow upwards along the faces of the indenter to the regions of the contact periphery. This mechanical process of piling up associated with the contact-induced plastic flows is essential, as it contributes considerably to the load-bearing capacity of the impression. A schematic of the enhanced plastic pile-up is depicted in Figure 2.17b for a soft film coated on a hard substrate pressed into contact with a cone/pyramid indenter at a load P. The purely elastic contact at the same indentation load P is also depicted in Figure 2.17a for comparison. As emphasized in Figure 2.17, due to the enhanced load-bearing capacity via the increased contact area through piling up, the total penetration depth h for the elastoplastic material may be smaller than that for a purely elastic indentation, although the elastoplastic contact depth h_c and the resultant contact area A_c are always larger than those for purely elastic contact.

Consider an unloading process at the maximum indentation load P. Similarly to the unloading process for an elastoplastic homogeneous infinite half-space, the penetration recovered during unloading in a layered composite is largely "elastic". This implies that the unloading stiffness S_{eff} at the maximum load P will also be efficient for estimating the elastic properties of a coating–substrate system. Chen and Vlassak conducted an elastoplastic FEA study of cone indentation for coating–substrate systems with various values of the ratio E'_f/E'_s and the ratio of yield strengths Y_f/Y_s [83]. They found that (1) the elastic mismatch, namely the value of modulus ratio E'_f/E'_s, plays a relatively minor role in the indentation hardness measurement of a coating film, the modulus ratio E'_f/E'_s being able to be neglected, in most cases, (2) the elastic mismatch has little importance if one determines the film hardness by directly measuring the contact area A_c and (3) the unloading stiffness S_{eff} is the same for both elastic and elastoplastic layered composites provided that the contact area A_c is the same, regardless of the facts that A_c and the degree of pile-up/sink-in significantly depend on Y_f/Y_s and that it is very different from that of an elastic contact with the same load (see Figure 2.17) [83]. These important findings suggest that the analytical procedures developed for the purely elastic contact problems of layered composites given in Section 2.5.1 can be successfully applied to the elastoplastic contact for estimating the contact area A_c and

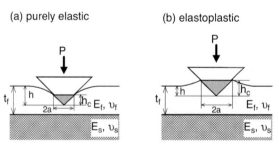

Figure 2.17 Significant effect of plastic flow on the impression profile of indentation, in which the geometric constraint of the substrate plays an essential role.

then determine the Meyer hardness H_M using the basic definition of the indentation hardness $H_M = P/A_c$ [see Eq. (20′)].

The unloading contact stiffness S_{eff} for an elastoplastic coating–substrate composite is given by the following expression after simply replacing E' with E'_{eff} in Eq. (27):

$$S_{eff} = \frac{2}{\sqrt{\pi}} \sqrt{A_c} \, E'_{eff} \tag{52}$$

Dividing both sides of Eq. (52) with $t_f E'_f$ and using the relation $A_c = \pi a^2$, the non-dimensional expression for the contact stiffness is given by

$$\frac{S_{eff}}{t_f E'_f} = 2 \left(\frac{a}{t_f} \right) \left(\frac{E'_{eff}}{E'_f} \right) \tag{53}$$

Accordingly, the analytical expression for E'_{eff}/E'_f given in Eq. (45) that is substituted into the right-hand side of Eq. (53) affords the analytical expression for the non-dimensional stiffness $S_{eff}/t_f E'_f$ as a function of the normalized contact radius a/t_f, shown in Figure 2.18 for various values of the modulus ratio E'_f/E'_s in conical indentation, as an example.

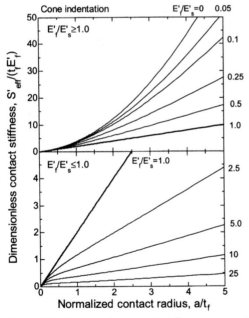

Figure 2.18 Dimensionless contact stiffness vs. normalized contact radius relations (cone/pyramid indentation) of coating–substrate composites for various values of the modulus ratio E'_f/E'_s. These characteristic lines are utilized for estimating the contact radius a from the observed unloading stiffness S_{eff} determined in indentation test (see Figure 2.19).

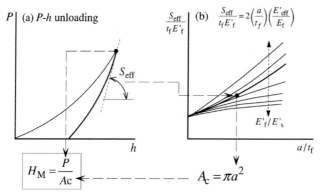

Figure 2.19 Experimental procedures for determining the Meyer hardness H_M of a coating–substrate system.

Once we have the analytical expression plotted in Figure 2.18 for the unloading contact stiffness S_{eff} in terms of a/t_f, we can determine the Meyer hardness of the coating film; the experimental procedure is depicted schematically in Figure 2.19. Figure 2.19a indicates the unloading stiffness S_{eff} at the maximum indentation load P. The S_{eff} value thus determined in the unloading process of micro/nanoindentation along with the values of the thickness t_f and the modulus E'_f of the film can then be used to determined the contact radius a and then the contact area A_c in Figure 2.19b (a schematic of Figure 2.18), *provided that we know both of the modulus values E'_f and E'_s*. Accordingly, using the definition of the Meyer hardness $H_M = P/A_c$, we can determine the Meyer hardness of the film. Prior to this experimental procedure, we need to know the values of E'_f and E'_s. If not, the iterative method shown in Figure 2.16 must be conducted by the use of a spherical indentation to determine the E'_f value. The experimental details of this procedure to determine the hardness of various types of thin films coated on various types of substrates are reported in the literature [95].

2.6
Conclusion

The indentation contact mechanics in the time-independent regime (elastic and elastoplastic deformations and flows) were discussed for homogeneous semi-infinite half-space and for coating–substrate systems. From the practical viewpoint, for characterizing the characteristic mechanical properties in micro/nanoindentation tests, the most critical problems and issues are the quantitative estimation of the contact area A_c, the situation being more crucial in coating–substrate systems, as the substrate significantly affects the pile-up/sink-in profile of hardness impression. As clearly discussed in this chapter, the approximations currently utilized for estimating A_c always include some critical assumptions and difficulties that

Figure 2.20 A bright-field image of the instrumented indentation microscope for a tetrahedral pyramid indentation on an aluminum alloy (A 5052) at $P = 59.5$ mN. The dark square at the center indicates the contact area A_c.

have not yet been overcome. A novel approach has recently been proposed by Miyajima and Sakai to determine directly the contact area A_c by the use of a newly developed indentation apparatus (instrumented indentation microscope) that is coupled with an optical system to measure "directly" the contact area through an optically transparent sapphire or diamond indenter [97]. This novel apparatus enables one to determine the mechanical properties (elastic, plastic and time-dependent viscoelastic properties) *without making any critical assumptions/approximations for estimating* A_c. An example of the contact view (bright-field image) is shown in Figure 2.20 for a tetrahedral diamond pyramid pressed into contact with an aluminum alloy (A5052) (indentation load $P = 59.5$ mN), where the dark square at the center indicates the contact area, A_c. Substituting the A_c value thus obtained with the present indentation microscope into Eqs. (20′) and (27) or Eq. (52), we can readily and precisely determine the Meyer hardness H_M and the elastic modulus E' or E'_{eff} of a homogeneous material or a coating–substrate composite without any critical assumptions/approximations for estimating the A_c value. This indentation microscope, in particular, has great potential for application to micro/ nanocomposites and coating–substrate layered materials.

2.7
Notation

a	Contact radius of axisymmetric impression; radius of flat-ended cylindrical punch
a_H	Contact radius of axisymmetric elastic impression for a homogeneous semi-infinite body
A_c	Projected contact area
A_r	Projected area of residual impression
C	Constraint factor in $H_M = CY$

C_{obs}	Elastic compliance of test system
C_f	Elastic compliance of test frame
D	Indentation ductility index defined by U_r/U_t
E	Young's modulus
E'	Elastic modulus of homogeneous body defined by $E/(1-v^2)$
E'_{app}	Apparent elastic modulus defined by $1/E'_{app} = 1/E' + 1/E'_i$
E'_{eff}	Effective elastic modulus of coating–substrate composite
E'_f	Elastic modulus of coating film
E'_i	Elastic modulus of indenter
E'_s	Elastic modulus of substrate
g	Geometric factor of cone/pyramid indenter defined by $A_c = gh_c^2$
$G(r, z; E, v)$	Boussinesq's Green function for point loading on a semi-infinite elastic half-space
h	Depth of penetration
h_c	Contact depth of penetration
h_e	Penetration depth in purely elastic indentation; elastic recovery in elastoplastic indentation
h_{max}	Depth of penetration at P_{max}
h_{bos}	Apparent penetration depth observed in indentation test
h_p	Penetration depth in purely plastic indentation; plastic depth of penetration in elastoplastic indentation
h_r	Residual depth of penetration
H	True hardness related to the yield strength Y through $H = CY$
H_M	Meyer hardness defined by P/A_r or P/A_c
k_1	Loading parameter defined by $P = k_1 h^2$ for cone/pyramid indentation
k_2	Unloading parameter defined by $P = k_2(h-h_r)^2$ for cone/pyramid indentation
k_e	Loading parameter in purely elastic indentation defined by $P = k_e h^2$ for cone/pyramid indentation
k_p	Loading parameter in purely plastic indentation defined by $P = k_p h^2$ for cone/pyramid indentation
m	Exponent in indentation unloading defined by $P = k_2(h-h_r)^m$
n	Exponent in indentation loading defined by $P = k_1 h^n$; $n = 1.0, 1.5$ and 2.0 for flat-ended punch, sphere and cone/pyramid indentations, respectively
$p(r; a)$	Contact stress distribution
$p_N(x)$	Normalized contact stress distribution defined by $p(r;a)/p_m$
p_m	Mean contact pressure defined by $P/\pi a^2$
P	Indentation load
P_{max}	Maximum indentation load
R	Radius of spherical indenter
S	Unloading stiffness of homogeneous body defined by dP/dh of the P–h unloading curve at P_{max}
S_{eff}	Unloading stiffness of coating–substrate composite

t_f	Thickness of coating film
U_e	Elastic energy released during indentation unloading defined by $\int_{h_r}^{h_{max}} P dh$ along the P–h unloading path
U_r	Hysteresis loop energy consumed for creating the residual hardness impression defined by U_t–U_e
U_t	Total energy (external work) applied to the material indented to the maximum penetration of h_{max} defined by $\int_0^{h_{max}} P dh$ along the P–h loading path
V_r	Volume of residual impression
$W(t_f/a)$	Stiffness weight function defined by $E'_{eff} = E'_s + (E'_f - E'_s)W(t_f/a)$
W_I	Work of indentation defined by U_r/V_r
Y	Yield strength of homogeneous body
Y_f	Yield strength of coating film
Y_s	Yield strength of substrate
Δh	Additional depth of penetration resulting from the elastic deformations and the mechanical clearances of test system
Δh_t	Tip truncation of cone/pyramid indenter
β	Inclined face angle of cone/pyramidal indenter
β_c	Inclined face angle of the equivalent cone which displaces the same volume for the pyramid indenter
β_{eff}	Effective face angle defined by β–β_r
β_r	Inclined face angle of residual impression
γ_1	Parameter of impression profile defined by h_{max}/h_c in elastoplastic indentation
γ_e	Parameter of impression profile defined by h_{max}/h_c in purely elastic indentation; 2 and $\pi/2$, respectively, for spherical and cone/pyramid indentations
γ_p	Parameter of impression profile defined by h_{max}/h_c in purely plastic indentation
$\bar{\varepsilon}$	Representative strain
$\bar{\sigma}$	Representative stress
ν	Poisson's ratio
ξ_r	Relative depth of residual impression defined by h_r/h_{max}

References

1 H. Hertz, *J. Reine Angew. Math.* **92** (1881) 156–171.

2 J. Boussinesq, *Application des Potentiels a l'Etude de l'Equilibre et du Movement des Solides Elastiques*, Gauthier-Villars, Paris 1885.

3 S. P. Timoshenko, J. N. Goodier, *Theory of Elasticity*, McGraw-Hill, New York 1951.

4 A. E. H. Love, *Quarterly J. Math.* **10** (1939) 161–168; *A Treatise on the Mathematical Theory of Elasticity*, Cambridge University Press, Cambridge 1952.

5 I. N. Sneddon, *Int. J. Eng. Sci.* **3** (1965) 47–57.

6 E. Meyer, *Z. Ver. Dtsch. Ing.* **52** (1908) 645–654.

7 L. Prandtl, *Nachr. Ges. Wiss. Göttingen*, (1920) 74–80.

8 R. Hill, *The Mathematical Theory of Plasticity*, Clarendon Press, Oxford 1950.

9 D. Tabor, *Hardness of Metals*, Clarendon Press, Oxford 1951.

10 K. L. Johnson, *Contact Mechanics*, Cambridge University Press, Cambridge 1985.

11 H. Staudinger, *Chem. Ber.* **53** (1920) 1073–1078.

12 H. Staudinger, *Angew. Chem.* **42** (1929) 67–73.

13 W. H. Carothers, *J. Am. Chem. Soc.* **51** (1929) 2548–2550.

14 P. J. Flory, *Principles of Polymer Chemistry*, Cornell University Press, Ithaca, NY 1953.

15 J. D. Ferry, *Viscoelastic Properties of Polymers*, Wiley, New York 1961.

16 J. R. M. Radok, *Q. Appl. Math.* **15** (1957) 198–202.

17 S. C. Hunter, *J. Mech. Phys. Solids*, **8** (1960) 219–234.

18 E. H. Lee, J. R. M. Radock, *J. Appl. Mech.* **27** (1960) 438–444.

19 T. C. T. Ting, *J. Appl. Mech.* **33** (1966) 845–854; **35** (1968) 248–254.

20 W. H. Yang, *J. Appl. Mech.* **33** (1966) 395–401.

21 S. Shimizu, T. Yanagimoto, M. Sakai, *J. Mater. Res.* **14** (1999) 4075–4086.

22 L. Cheng, X. Xia, W. Yu, L. E. Scriven, W. W. Gerberich, *J. Polym. Sci., Part B: Polym. Phys.* **38** (2000) 10–22.

23 M. Sakai, S. Shimizu, N. Miyajima, Y. Tanabe, E. Yasuda, *Carbon* **39** (2001) 605–614.

24 M. Sakai, S. Shimizu, *J. Non-Cryst. Solids* **282** (2001) 236–247.

25 M. Sakai, S. Shimizu, *J. Am. Ceram. Soc.* **85** (2002) 1210–1216.

26 M. Sakai, *Philos. Mag. A* **82** (2002) 1841–1849.

27 N. A. Stilwell, D. Tabor, *Proc. Phys. Soc. London*, **78** (1961) 169–179.

28 B. R. Lawn, V. R. Howes, *J. Mater. Sci.* **16** (1981) 2745–2752.

29 M. Kh. Shorshorov, S. I. Bulychev, V. A. Alekhim, *Sov. Phys. Dokl.* **26** (1981) 769–771.

30 D. Newey, M. A. Wilkins, H. M. Pollock, *J. Phys. E: Sci. Instrum.* **15** (1982) 119–122.

31 F. Frölich, P. Grau, W. Grellmann, *Phys. Status Solidi A* **42** (1977) 79–80.

32 J. B. Pethica, R. Hutchings, W. C. Oliver, *Philos. Mag. A* **48** (1983) 593–606.

33 J. L. Loubet, J. M. Georges, O. Marchesini, G. Meille, *J. Tribol.* **106** (1984) 43–48.

34 J. L. Loubet, J. M. Georges, G. Meille, Microindentation, In P. J. Blau, B. R. Lawn (eds.), *Techniques in Materials Science and Engineering*, ASTM STP889, 72–89, American Society for Testing and Materials, Philadelphia 1986.

35 M. F. Doerner, W. D. Nix, *J. Mater. Res.* **1** (1986) 601–609.

36 D. Stone, W. R. LaFontaine, P. Alexopoulos, F.-W. Wu, C.-Y. Li, *J. Mater. Res.* **3** (1988) 141–147.

37 D. L. Joslin, W. C. Oliver, *J. Mater. Res.* **5** (1990) 123–126.

38 G. M. Pharr, R. F. Cook, *J. Mater. Res.* **5** (1990) 847–851.

39 T. F. Page, W. C. Oliver, C. J. McHargue, *J. Mater. Res.* **7** (1992) 450–475.

40 G. M. Pharr, W. C. Oliver, F. R. Brotzen, *J. Mater. Res.* **7** (1992) 613–617.

41 W. C. Oliver, G. M. Pharr, *J. Mater. Res.* **7** (1992) 1564–1583.

42 M. Sakai, *Acta Metall. Mater.* **41** (1993) 1751–1758.

43 J. S. Field, M. V. Swain, *J. Mater. Res.* **8** (1993) 297–306.

44 R. Nowak, M. Sakai, *J. Mater. Res.* **8** (1993) 1068–1078.

45 E. Söderlund, D. J. Rowcliffe, *J. Hard Mater.* **5** (1994) 149–177.

46 R. F. Cook, G. M. Pharr, *J. Hard Mater.* **5** (1994) 179–190.

47 K. Zeng, E. Söderlund, A. E. Giannakopoulos, D. J. Rowcliffe, *Acta Mater.* **44** (1996) 1127–1141.

48 S. V. Hainsworth, H. W. Chandler, T. F. Page, *J. Mater. Res.* **11** (1996) 1987–1994.

49 Y.-T. Cheng, C.-M. Cheng, *J. Appl. Phys.* **84** (1998) 1284–1291.

50 Y.-T. Cheng, C.-M. Cheng, *Int. J. Solids Struct.* **36** (1999) 1231–1243.

51 M. Sakai, S. Shimizu, T. Ishikawa, *J. Mater. Res.* **14** (1999) 1471–1484.

52 A. C. Fischer-Cripps, *Vacuum* **58** (2000) 569–585.

53 M. Sakai, Y. Nakano, *J. Mater. Res.* **17** (2002) 2161–2173.

54 G. M. Pharr, A. Bolshakov, *J. Mater. Res.* **17** (2002) 2660–2671.

55 M. Sakai, *J. Mater. Res.* **18** (2003) 1631–1640.

56 W. C. Oliver, G. M. Pharr, *J. Mater. Res.* **19** (2004) 3–20.

57 Y.-T. Cheng, C.-M. Cheng, *Mater. Sci. Eng. R* **44** (2004) 91–149.

58 A. Simono, K. Tanaka, Y. Akiyama, H. Yoshizaki, *Philos. Mag. A* **74** (1996) 1097–1105; T. Sawa, K. Tanaka, *Philos. Mag. A* **82** (2002) 1851–1856.

59 M. Sakai, H. Hanyu, M. Inagaki, *J. Am. Ceram. Soc.* **78** (1995) 1006–1012.

60 M. Sakai, T. Akatsu, S. Numata, K. Matsuda, *J. Mater. Res.* **18** (2003) 2087–2096.

61 J. Zhang, M. Sakai, *Mater. Sci. Eng. A* **381** (2004) 62–70.

62 M. Sakai, *J. Mater. Res.* **14** (1999) 3630–3639.

63 M. Sakai, T. Akatsu, S. Numata, *Acta Mater.* **52** (2004) 2359–2364.

64 N. Hakiri, M. Sakai, unpublished data.

65 M. Sakai, R. Nowak, in M. J. Bannister (ed.), *Ceramics, Adding the Value*, Vol. 2, 922–931, Australian Ceramic Society, Melbourne 1992.

66 R. S. Dhaliwal, *Int. J. Eng. Sci.* **8** (1970) 273–288.

67 R. S. Dhaliwal, I. S. Rau, *Int. J. Eng. Sci.* **8** (1970) 843–856.

68 W. T. Chen, P. A. Engel, *Int. J. Solids Struct.* **8** (1972) 1257–1281.

69 R. B. King, *Int. J. Solids Struct.* **23** (1987) 1657–1664.

70 H. Y. Yu, S. C. Sanday, B. B. Rath, *Mech. Phys. Solids* **38** (1990) 745–764.

71 T. W. Wu, *J. Mater. Res.* **6** (1991) 407–426.

72 H. Gao, C.-H. Chiu, J. Lee, *Int. J. Solids Struct.* **29** (1992) 2471–2492.

73 H. Gao, T. W. Wu, *J. Mater. Res.* **8** (1993) 3229–3232.

74 W. W. Gerberich, J. T. Wyrobek, *Acta Mater.* **44** (1996) 3585–3598.

75 D. F. Bahr, J. W. Hoehm, N. R. Moody, W. W. Gerberich, *Acta Mater.* **45** (1997) 5163–5175.

76 J. Menčík, D. Munz, E. Quandt, E. R. Weppelmann, *J. Mater. Res.* **12** (1997) 2475–2484.

77 N. R. Moody, R. Q. Hwang, S. V. Taraman, J. E. Angelo, D. P. Norwood, W. W. Gerberich, *Acta Mater.* **46** (1998) 585–597.

78 J. Malzbender, G. With, *J. Non-Cryst. Solids* **265** (2000) 51–60.

79 J. Malzbender, G. With, *Surf. Coat. Technol.* **135** (2000) 60–68.

80 J. Malzbender, G. With, J. Toonder, *Thin Solid Films* **372** (2000) 134–143.

81 J. Malzbender, G. With, J. Toonder, *Thin Solid Films* **366** (2000) 139–149.

82 A. Gouldstone, H.-J. Koh, K.-Y. Zeng, A. E. Giannakopoulos, S. Suresh, *Acta Mater.* **48** (2000) 2277–2293.

83 X. Chen, J. J. Vlassak, *J. Mater. Res.* **16** (2001) 2974–2982.

84 J. Toonder, J. Malzbender, G. With, R. Balkenende, *J. Mater. Res.* **17** (2002) 224–233.

85 R. Saha, W. D. Nix, *Acta Mater.* **50** (2002) 23–38.

86 J. Malzbender, G. With, *Surf. Coat. Technol.* **154** (2002) 21–26.

87 F. Yang, *Mater. Sci. Eng. A* **358** (2003) 226–232.

88 T. Y. Tsui, C. A. Ross, G. M. Pharr, *J. Mater. Res.* **18** (2003) 1383–1391.

89 C.-H. Hsueh, P. Miranda, *J. Mater. Res.* **19** (2004) 94–100.

90 C.-H. Hsueh, P. Miranda, *J. Mater. Res.* **19** (2004) 2774–2781.

91 Y.-G. Jung, B. R. Lawn, M. Martyniuk, H. Huang, X. Z. Hu, *J. Mater. Res.* **19** (2004) 3076–3080.

92 A. J. Atanacio, B. A. Latella, C. J. Barbé, M. V. Swain, *Surf. Coat. Technol.* **192** (2005) 354–364.

93 M. Sakai, J. Zhang, A. Matsuda, *J. Mater. Res.* **20** (2005) 2173–2183.

94 M. Sakai, M. Sasaki, A. Matsuda, *Acta Mater.* **53** (2005) 4455–4462.

95 S. M. Han, R. Saha, W. D. Nix, *Acta Mater.* **54** (2006) 1571–1581.

96 M. Sakai, *Philos. Mag. A* **86** (2006) 5607–5624.

97 T. Miyajima, M. Sakai, *Philos. Mag. A* **86** (2006) 5729–5737; M. Sakai, N. Hakiri, T. Miyajima, *J. Mater. Res.* **31** (2006) 2298–2303.

3
Thin-film Characterization Using the Bulge Test

Oliver Paul and Joao Gaspar, Department of Microsystems Engineering (IMTEK), University of Freiburg, Germany

Abstract

The bulge test, i.e., the load-deflection response of thin films under differential pressure, has evolved into a mature method for the mechanical characterization of thin film materials. The chapter reviews the state-of-the-art in modelling and characterizing uniformly loaded diaphragms and on the extraction of elastic properties such as Young's modulus and Poisson's ratio, prestress and prestrain, coefficient of thermal expansion, and fracture-mechanical parameters such as fracture strength and Weibull modulus from such structures. Powerful models for square and long rectangular thin plates are available. The models for long membranes are able to handle multilayer structures and elastic supports. A novel setup enabling the characterization of up to 80 membranes on a four-inch wafer in a single measurement run is reported. Mechanical property data extracted using the bulge test from thin films made of silicon nitrides, silicon oxides, polysilicon, metals, polymers and other materials relevant for microelectromechanical systems are reviewed.

Keywords

bulge test; thin films; diaphragms; membranes; MEMS; thin plates; load-deflection; Young's modulus; elastic modulus; mechanical properties; Poisson's ratio; prestress; residual stress; multilayer; fracture strength; Weibull modulus

Reliability of MEMS.
Edited by O. Tabata, T. Tsuchiya
Copyright © 2013 WILEY-VCH Verlag GmbH & Co. KGaA, Weinheim
ISBN: 978-3-527-33501-5

3.1
Introduction

In the bulge test, the mechanical parameters of thin films are extracted from the load-deflection response of thin-film diaphragms under varying differential pressure p. The thin film is laterally constrained by a substrate or a sample holder. The method was first introduced by Beams [1] long before the inception of microelectromechanical systems (MEMS), where thin films play a fundamental role as mechanical materials. In that work, polycrystalline and single-crystal gold and polycrystalline silver films were investigated. Using a sophisticated sample preparation method, the films were deposited over a circular orifice. Approximate formulae for the mechanical response of structures consisting of isotropic and anisotropic materials were provided and compared with experimental results [2]. For the circular geometry, elaborate mechanical models are now available [3].

In comparison with this pioneering work, sample preparation was considerably simplified by the emergence of micromachining techniques combined with thin-film deposition borrowed from IC technology. Thin films were deposited either on both sides of a (100) silicon wafer, or on one side only, whereas the rear face of the wafer was covered by a structurable etch mask layer. Rear etching in appropriate anisotropic etchants such as potassium hydroxide (KOH) or ethylenedi-amine–pyrocatechol (EDP) solutions led to the formation of etch cavities with well-defined sidewalls belonging to the {111} family of crystal planes. For cavities of sufficiently large lateral dimensions, a well-defined square or rectangular membrane consisting of the material of interest on the front face of the wafer remained suspended on a silicon substrate frame [4, 5].

A structure of this type is shown schematically in Figure 3.1. The thin-film diaphragm has a thickness t and side lengths a and b. The load-deflection response of such a structure is governed by essentially three effects: first, the tendency of the perpendicular load to drive the structure away from its load-free equilibrium position; second, the counteraction by the in-plane stresses against the deformation; third, the resistance opposed by the bending stiffness of the thin film against the deformation. If the diaphragm is very thin, the influence of bending is negligible. In the language of mechanics, the structure is then termed a *membrane*. If, in contrast, bending stiffness plays a role but the material is still thin enough to satisfy the Kirchhoff condition of negligible out-of-plane shear stresses, the structure is termed a *thin plate* [7, 8]. The equilibrium conditions governing the load-deflection response of thin plates, the corresponding simpler form for membranes and their consequences are developed in the following.

The deflection profile of the membrane under the load p is determined by the elastic modulus or Young's modulus E of the membrane material, its Poisson's ratio ν and its prestress σ_0 for isotropic materials. These are the properties that one may in principle hope to extract from load-deflection responses of thin-film diaphragms. The experimental freedom provided by the geometric parameters t, a and b, the possibility to build up multilayers to characterize even samples that may prove difficult to assess as stand-alone materials and the relative robustness

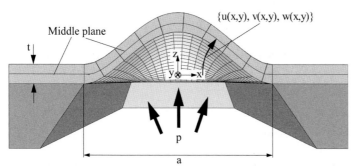

Figure 3.1 Schematic view of the bulge test, i.e. the deflection of a thin-film diaphragm under a differential pressure load p [6]. In comparison with experimental reality, the displacement is strongly exaggerated.

of the samples partly explains the popularity of the bulge test for determining mechanical thin-film properties.

Among researchers active in the field, major contributions were made by the groups of

- S. D. Senturia [9–13]
- O. Tabata [14–18]
- W. D. Nix [19–22]
- E. Obermeier [23–26]
- A. Bosseboeuf [27–30] and
- the authors of this chapter [31–42].

A broad variety of materials have been investigated using the bulge test. Among these are materials typically used in the semiconductor industry and others, including

- silicon [43–45]
- polysilicon by low-pressure chemical vapor deposition (LPCVD) [14, 15, 17, 24–26, 46–50]
- silicon oxide [33, 44, 48, 49, 51–55]
- silicon oxynitride [29, 30]
- silicon nitride from LPCVD and plasma-enhanced chemical vapor deposition (PECVD) [5, 6, 14, 15, 18, 22, 24, 26, 31, 33–40, 48, 49, 56–62]
- titanium nitride [63]
- gold [1, 5, 64]
- silver [1]
- tungsten [5, 27, 65]
- aluminum [5, 49, 65]
- copper [2, 66]
- molybdenum [28]
- nickel–iron permalloy [67]
- silicon carbide [58, 68, 69]
- zinc oxide [56, 59] and
- polymer thin films [2, 9, 10–12, 70–74].

In order to approach this topic systematically, we will start in Section 3.2 by laying out the fundamentals for building up models of thin-film diaphragms for the reliable extraction of mechanical parameters. In Section 3.3, currently available load-deflection models of thin-film diaphragms are described and analyzed. Experimental developments and issues are discussed in Section 3.4. Mechanical property data extracted using the bulge test are summarized in Section 3.5 and the conclusions in Section 3.6 close the chapter.

3.2
Theory

This section develops the basic mechanics of the load-deflection response of thin plates. We will first formulate the governing equations for single-material diaphragms in Section 3.2.1. In Section 3.2.2, these are then extended to the case of long multilayer diaphrams, under plane-strain deformations. In order to elucidate the number of independent parameters governing the diaphragm mechanics and to simplify the corresponding equations, the results of these two sections are cast into dimensionless form in Section 3.2.3. The corresponding dimensionless mechanics of membranes is described in Section 3.2.4. Finally, Section 3.2.5 summarizes the general procedure that enables to extract the mechanical parameters of diaphragms.

3.2.1
Basic Definitions for Single-layer Diaphragms

The starting assumption of this section is that the diaphragm consists of a single isotropic material with elastic modulus E and Poisson's ratio v, and prestrained to ε_0. The deformation of a thin plate under the presssure p is described by the vector field of the displacements $\{u(r),\ v(r),\ w(r)\}$ of the individual points $r = \{x, y\}$ of its middle plane (the neutral fiber in classical plate theory) in the two horizontal in-plane and the out-of-plane directions, as shown in Figure 3.1. These displacements result in the strain tensor components ε_{xx}, ε_{yy} and ε_{xy} including the residual strain ε_0:

$$\varepsilon_{xx} \equiv \varepsilon_0 + \varepsilon_{1,xx} = \varepsilon_0 + \frac{\partial}{\partial x} u(r) + \frac{1}{2}\left[\frac{\partial}{\partial x} w(r)\right]^2$$

$$\varepsilon_{yy} \equiv \varepsilon_0 + \varepsilon_{1,yy} = \varepsilon_0 + \frac{\partial}{\partial y} v(r) + \frac{1}{2}\left[\frac{\partial}{\partial y} w(r)\right]^2$$

$$\varepsilon_{xy} = \varepsilon_{yx} \equiv \varepsilon_{1,xy} = \frac{1}{2}\left[\frac{\partial}{\partial y} u(r) + \frac{\partial}{\partial x} v(r)\right] + \frac{1}{2}\left[\frac{\partial}{\partial x} w(r)\right]\left[\frac{\partial}{\partial y} w(r)\right] \tag{1}$$

which define $\varepsilon_{1,xx}$, $\varepsilon_{1,yy}$ and $\varepsilon_{1,xy} = \varepsilon_{1,yx}$ as the strain components caused by the external load adding to the residual strain ε_0 present before the load is applied. These total strains are translated into the in-plane stress tensor components in the mid-plane of the structure by the stress–strain relationships of thin plates [6, 75]:

$$\sigma_{xx} = \sigma_0 + \frac{E}{(1-v^2)}(\varepsilon_{1,xx} + v\varepsilon_{1,yy}) \equiv \sigma_0 + \sigma_{1,xx}$$

$$\sigma_{yy} = \sigma_0 + \frac{E}{(1-v^2)}(\varepsilon_{1,yy} + v\varepsilon_{1,xx}) \equiv \sigma_0 + \sigma_{1,yy}$$

$$\sigma_{xy} = \sigma_{yx} = \frac{E}{(1+v)}\varepsilon_{1,xy} \equiv \sigma_{1,xy} \tag{2}$$

where $\sigma_0 = \varepsilon_0 E/(1 - v)$ denotes the residual stress of the thin structure. In view of the biaxial character of the residual stress, $E/(1 - v)$ is termed biaxial modulus.

The conditions governing the out-of-plane and in-plane equilibrium of such a structure are then [76]

$$D\Delta^2 w(r) - t\sigma_0 \Delta w(r) - t \sum_{x_i, x_j = x, y} \left[\sigma_{1, x_i x_j}(r) \frac{\partial^2}{\partial x_i \partial x_j} w(r) \right] = p \tag{3}$$

$$\sum_{x_j = x, y} \frac{\partial}{\partial x_j} \sigma_{1, x_i x_j}(r) = 0 \quad \text{for} \quad x_i = x, y \tag{4}$$

respectively, where $\Delta = \partial^2/\partial x^2 + \partial^2/\partial y^2$ denotes the two-dimensional Laplace operator and $D = Et^3/12(1 - v^2)$ is the flexural rigidity of the diaphragm. The first, second and third terms on the left-hand side of Eq. (3) correspond to the restoring force due to out-of-plane bending, residual membrane forces and the additional membrane forces due to the additional strain in the structure under load, respectively. The restoring forces have to match the applied pressure load p. Equations (3) and (4) have to be solved for u, v and w subject to the boundary conditions

$$u(r_b) = v(r_b) = w(r_b) = 0 \tag{5}$$

and [32, 77]

$$D \frac{\partial^2}{\partial x_\perp^2} w(r_b) = -K \frac{\partial}{\partial x_\perp} w(r_b) \tag{6}$$

where r_b and $\partial/\partial x_\perp$ denote the coordinates of any point of the diaphragm boundary and the derivative normal to the boundary in the outward direction, respectively. Equation (6) implements the assumption of an elastic membrane support, relating the torque exerted by the diaphragm on the support to its slope at the boundary. The rotational spring parameter K has been estimated using finite-element (FE) simulations as $K = t^2 E_{ps}^{(substr.)}/k_r$, where $E_{ps}^{(substr.)} = E_{substr.}/(1 - v_{substr.}^2)$ and k_r denote the plane-strain modulus of the substrate and a dimensionless geometry factor [77], respectively. For wet-etched membranes with {111} oriented cavity wall with 54.7° inclination, k_r was found to be roughly 0.78.

For elastic supports, one may consider relaxing the boundary conditions in Eq. (5) [77]. Displacements in the horizontal and vertical directions are possible and to lowest order are proportional to the in-plane and out-of-plane loads, respectively. However, for all practical purposes in bulge testing in the field of MEMS, the effect of these boundary displacements was found by the authors to be negligible.

In the case of the long membranes treated in more detail in Sections 3.3.4–3.3.6, the diaphragms undergo a plane-strain deformation. The term *plane strain* means that application of the pressure displaces membrane points only in xz planes and leaves the y coordinates unchanged. The out-of-plane equilibrium condition is reduced to a differential equation in one variable, namely

$$D \frac{d^4 w}{dx^4} - t\sigma_{xx} \frac{d^2 w}{dx^2} = p \tag{7}$$

Similarly, the in-plane condition reduces to $d\sigma_{xx}/dx = 0$. By considering the first equality of Eq. (2) and replacing ε_{xx} by the expression given in Eq. (1), integrating from $-a/2$ to $a/2$, and noting that the integral of du/dx vanishes in view of the boundary condition of Eq. (5), we obtain

$$\sigma_{xx} = \sigma_0 + \frac{E_{ps}}{2a} \int_{-a/2}^{a/2} \left(\frac{dw}{dx}\right)^2 dx \tag{8}$$

This equation describes the following fact: due to the deflection, the diaphragm undergoes a relative length change represented by the integral divided by $2a$; via the plane-strain modulus $E_{ps} = E/(1 - v^2)$, this length change results in an additional stress adding to the initial stress σ_0. Equation (8) in conjunction with Eq. (7) represent the non-linear response of elastically supported plane-strain membranes under uniform pressure load. The response of the membrane is non-linear because of the geometric non-linearity appearing as the last term in Eq. (8).

3.2.2
Multilayers Under Plane-strain Deformations

In multilayer diaphragms, the situation is further complicated, for several reasons. First, the individual layers have specific thicknesses t_n, elastic properties, i.e. elastic moduli E_n and Poisson's ratios v_n, and are subjected to individual isotropic prestrains and prestresses $\varepsilon_{0,n}$ and $\sigma_{0,n}$, respectively, where the index $n = 1, 2, \ldots$, N enumerates the component layers. In more general terms, this may be stated as a z-dependent elastic modulus $E(z)$, Poisson's ratio $v(z)$, prestrain $\varepsilon_0(z)$, prestress $\sigma_0(z) = E(z)\varepsilon_0(z)/[1 - v(z)]$, additional stress $\sigma_{1,xx}(z)$ due to the deformation under pressure and total membrane stress $\sigma_{xx}(z)$. In plane-strain situations, then, the z-dependent plane-strain modulus $E_{ps}(z) = E(z)/[1 - v^2(z)]$ plays a central role.

The second point is that the mid-plane loses its significance as the neutral plane. In fact, depending on the multilayer composition, the mid-plane will in general experience an in-plane stress under bending and therefore will not necessarily be neutral. However, it is useful to define alternatively a reference plane z_0 by the condition stated in Table 3.1. In the integral, the bottom and top surfaces of the diaphragm are assumed to be at the coordinates $z = 0$ and $z = t$. This definition of z_0 is equivalent to the request that the bending moment due to in-plane plane-strain elongation ($\varepsilon_{1,xx} \neq 0$, $\varepsilon_{1,yy} = 0$) of the membrane is zero with respect to z_0, i.e.

$$\int_0^t \sigma_{1,xx}(z)(z - z_0)dz = \varepsilon_{1,xx} \int_0^t E_{ps}(z)(z - z_0)dz = 0 \tag{9}$$

This assumption ensures in the decoupling of energy terms corresponding to elongation and bending of the diaphragm under plane-strain deformations, as is naturally the case with single-layer membranes with the standard definition of the neutral plane.

Table 3.1 Integral parameters for long multilayer membranes under plane-strain conditions and their explicit form. The constant k_r has been found to be about 0.78 for diaphragms defined on bulk-micromachined Si frames under uniform differential pressure loads [77].

Parameter	Integral form	Explicit form	Single-layer form
Planestrain modulus [N m^{-2}]	E_{ps} $\dfrac{E(z)}{1-v^2(z)}$	$\dfrac{E_n}{1-v_n^2}$	$\dfrac{E}{1-v^2}$
Reference plane [m]	z_0 $\dfrac{\int_0^t E_{ps}(z)z\,dz}{\int_0^t E_{ps}(z)\,dz}$	$\dfrac{\sum_{n=1}^N\left[E_{ps,n}t_n\left(\sum_{m=1}^n t_m - t_n/2\right)\right]}{\sum_{n=1}^N E_{ps,n}t_n}$	$\dfrac{t}{2}$
Stretching stiffness [N m]	D_0 $\int_0^t E_{ps}(z)\,dz$	$\sum_{n=1}^N E_{ps,n}t_n$	$E_{ps}t$
Bending stiffness [N m]	D_2 $\int_0^t E_{ps}(z)(z-z_0)^2\,dz$	$\sum_{n=1}^N\left\{E_{ps,n}t_n\prod_{l=1}^2\left[\sum_{m=1}^n t_m -(1+i(-1)^l/\sqrt3)t_n/2 - z_0\right]\right\}$	$\dfrac{E_{ps}t^3}{12}$
Initial line force [N m]	S_0 $\int_0^t \sigma_0(z)\,dz$	$\sum_{n=1}^N \sigma_{0,n}t_n$	$\sigma_0 t$
Total line force [N m]	S $\int_0^t \sigma_{xx}(z)\,dz$	–	$\sigma_{xx}t$
Initial bending moment/ length [N]	S_1 $\int_0^t \sigma_0(z)(z-z_0)\,dz$	$\sum_{n=1}^N\left[\sigma_{0,n}t_n\left(\sum_{m=1}^n t_m - t_n/2 - z_0\right)\right]$	0
Stiffness of supports [N]	K $\dfrac{E_{ps}^{(substr.)}t^2}{k_r}$	$\dfrac{E_{ps}^{(substr.)}\left(\sum_{n=1}^N t_n\right)^2/k_r}{k_r}$	$\dfrac{E_{ps}^{(substr.)}t^2}{k_r}$

Third, the roles of E_{ps}, σ_0, and σ_{xx} in Eq. (8) are taken over by the elastic modulus, prestress and total membrane stress integrated over the thickness of the multilayer, i.e. via the stretching stiffness D_0, the initial line force S_0 and the total line force S, as defined in Table 3.1. Further, the resistance of the multilayer against plane-strain bending is described by the parameter D_2 as defined in Table 3.1. Morever, due to the prestress distribution, a cross-section of the multilayer is always exposed to an initial bending moment per unit length of the diaphragm, S_1. This effect is absent in single-layer materials. Nonzero values of S_1 would lead to a curvature of a piece of the multilayer with free edges. Table 3.1 shows how S_1 is computed in terms of the prestress distribution in the multilayer. Finally, the K value for the support stiffness is taken over from the single-material case, with the total thickness t of the multilayer playing the role of the thickness parameter.

Table 3.1 also lists explicit expressions for the parameters $E_{ps,n}$, z_0, D_0, D_2, S_0, S_1 and K for the case where the multilayer is composed of a stack of well-defined materials, each with individual thicknesses t_n and z-independent mechanical properties. The integrals for the continuously varying properties can then be written explicitly as sums of appropriate terms. In addition, Table 3.1 lists the expressions into which these lengthy equations simplify in the case of single-layer diaphragms.

With these definitions, Eqs. (7) and (8) read

$$D_2 \frac{d^4 w}{dx^4} - S \frac{d^2 w}{dx^2} = p \tag{10}$$

$$S = S_0 + \frac{D_0}{2a} \int_{-a/2}^{a/2} \left(\frac{dw}{dx}\right)^2 dx \tag{11}$$

Boundary condition Eq. (5) reduces to

$$u(\pm a/2) = w(\pm a/2) = 0 \tag{12}$$

while boundary condition Eq. (6) becomes

$$D_2 \frac{d^2 w}{dx^2}\bigg|_{\pm a/2} \pm K \frac{dw}{dx}\bigg|_{\pm a/2} = S_1 \tag{13}$$

In summary, these equations describe the non-linear load-deflection behavior of elastically clamped prestressed thin multilayer plates under plane strain.

3.2.3
Reduction to Dimensionless Form

Considering Eqs. (1)–(6), one might expect the load-deflection response of a single-layer rectangular diaphragm to depend on the geometry parameters a, b and t, on the mechanical parameters ε_0, E, v and K and on the load p. Similarly, in view of Eqs. (10)–(13), the load-deflection behavior of multilayer diaphragms under plane-strain conditions depends on the parameters a, D_0, D_2, S_0, S_1, K and p. However, this impressive number of seemingly independent parameters is effectively reduced to a more modest set on the basis of a dimensional analysis based on the following definitions of dimensionless parameters:

- Coordinate: $\bar{x} = x/a$
- Deflection: $\bar{w} = w\sqrt{D_0/D_2}$
- Pressure: $\bar{p} = pa^4 \sqrt{D_0/D_2^3}$
- Effective line force: $\bar{S} = Sa^2/D_2$
- Initial line force: $\bar{S_0} = S_0 a^2/D_2$
- Initial bending moment per unit length: $\bar{S_1} = S_1 a^2 \sqrt{D_0/D_2^3}$
- Stiffness of the supporting edges: $\bar{K} = Ka/D_2$

In addition to the perspective of simplified equations, the main reason for going through such a renormalization exercise is that it reveals the independent pieces of information that one may hope to extract from load-deflection measurements of diaphragms.

With the above definitions, the equilibrium conditions for a multilayer membrane under plane strain, i.e. Eqs. (10) and (11) supplemented by the boundary conditions, i.e. Eqs. (12) and (13), become

$$\frac{d^4}{d\bar{x}^4}\bar{w} - \bar{S}\frac{d^2}{d\bar{x}^2}\bar{w} = \bar{p} \tag{14}$$

$$\bar{S} = \bar{S}_0 + \frac{1}{2}\int_{-1/2}^{1/2}\left(\frac{d}{d\bar{x}}\bar{w}\right)^2 d\bar{x} \tag{15}$$

$$\bar{u}\left(\pm\frac{1}{2}\right) = \bar{w}\left(\pm\frac{1}{2}\right) = 0 \tag{16}$$

and

$$\frac{d^2}{d\bar{x}^2}\bar{w}\bigg|_{\pm 1/2} \pm \bar{K}\frac{d}{d\bar{x}}\bar{w}\bigg|_{\pm 1/2} = \bar{S}_1 \tag{17}$$

It is evident that the load-deflection response depends on only four independent parameters: the load $\bar{p}$, the reduced residual stress parameter $\bar{S}_0$, the rotational suspension stiffness $\bar{K}$ and the line bending moment parameter $\bar{S}_1$ related to inhomogeneous stress across the layer sandwich. The solution to Eqs. (10)–(13) therefore must have the compact form $\bar{w}(\bar{x}, \bar{p}) = f_{ps}(\bar{x}, \bar{p}, \bar{S}_0, \bar{K}, \bar{S}_1)$. As a consequence, expressed in terms of the real, physical quantities x, p, S_0, K and S_1, the solution has the structure

$$w(x, p) = \sqrt{\frac{D_2}{D_0}} \times f_{ps}\left(\frac{x}{a}, p\frac{a^4}{D_2}\sqrt{\frac{D_0}{D_2}}, S_0\frac{a^2}{D_2}, K\frac{a}{D_2}, S_1\frac{a^2}{D_2}\sqrt{\frac{D_0}{D_2}}\right) \tag{18}$$

In order to describe all plane-strain load-deflection problems of elastically supported, long multilayer diaphragms, it is therefore sufficient to find and analyze the function f_{ps}. This is done in Sections 3.3.4–3.3.6. Equation (18) shows, for instance, that it is impossible to extract Poisson's ratio from plane-strain bulge measurements, since f_{ps} does not explicitly depend on v. All one may hope to extract from samples with difference dimensions a is S_0, i.e. the average residual stress, D_0, i.e. the average plane-strain modulus, D_2, i.e. the flexural rigidity, S_1, i.e., the residual bending moment, and K, if the latter is not assumed anyway as a numerical input based on its definition in Table 3.1.

In the more general case of Eqs. (3)–(6) applicable to single-layer diaphragms, a similar dimensionless form of the solution can be constructed. This requires the introduction of additional renormalized quantities, namely $\bar{r} = r/a$, i.e. $\bar{x} = x/a$, $\bar{y} = y/a$ and consequently $\bar{r}_b = r_b/a$, $\bar{x}_\perp = x_\perp/a$ and $\bar{\Delta} = a^2\Delta$. With the definitions $\bar{u} = ua D_0/D_2$, $\bar{v} = va D_0/D_2$, and $\bar{\varepsilon}_0 = \varepsilon_0 a^2 D_0/D_2$ and their insertion into Eq. (1), one obtains $\bar{\varepsilon}_{ij} = \varepsilon_{ij}a^2 D_0/D_2$ and $\bar{\varepsilon}_{1,ij} = \varepsilon_{1,ij}a^2 D_0/D_2$ for $i,j = x,y$. These definitions and $\bar{s}_0 = \sigma_0 a^2 t/D_2 = \bar{S}_0$ result in $\bar{s}_{ij} = \sigma_{ij}a^2 t/D_2$ and thus $\bar{s}_{1,ij} = \sigma_{1,ij}a^2 t/D_2$; finally, we define $\bar{p} = pa^4 D_0^{1/2}/D_2^{3/2}$ and $\bar{K} = Ka/D_2 = at^2 E_{ps}^{(substr.)}/k_r D_2$.

With these rules and $\bar{s}_0 = \bar{\varepsilon}_0(1+v)$, the set of definitions in Eq. (1) is translated into

$$\overline{\varepsilon_{xx}} \equiv \bar{\varepsilon}_0 + \overline{\varepsilon_{1,xx}} = \bar{\varepsilon}_0 + \frac{\partial}{\partial \bar{x}}\bar{u} + \frac{1}{2}\left(\frac{\partial}{\partial \bar{x}}\bar{w}\right)^2$$

$$\overline{\varepsilon_{yy}} \equiv \bar{\varepsilon}_0 + \overline{\varepsilon_{1,yy}} = \bar{\varepsilon}_0 + \frac{\partial}{\partial \bar{y}}\bar{v} + \frac{1}{2}\left(\frac{\partial}{\partial \bar{y}}\bar{w}\right)^2$$

and

$$\overline{\varepsilon_{xy}} = \overline{\varepsilon_{yx}} \equiv \overline{\varepsilon_{1,xy}} = \frac{1}{2}\left(\frac{\partial}{\partial \bar{y}}\bar{u} + \frac{\partial}{\partial \bar{x}}\bar{v}\right) + \frac{1}{2}\left(\frac{\partial}{\partial \bar{x}}\bar{w}\right)\left(\frac{\partial}{\partial \bar{y}}\bar{w}\right) \tag{19}$$

As a consequence, Eq. (2) simplifies to

$$\overline{s_{xx}} = \bar{\varepsilon}_0(1+v) + \overline{\varepsilon_{1,xx}} + v\overline{\varepsilon_{1,yy}} = \bar{\varepsilon}_0(1+v) + \overline{s_{1,xx}}$$

$$\overline{s_{yy}} = \bar{\varepsilon}_0(1+v) + \overline{\varepsilon_{1,yy}} + v\overline{\varepsilon_{1,xx}} = \bar{\varepsilon}_0(1+v) + \overline{s_{1,yy}}$$

$$\overline{s_{xy}} = \overline{s_{yx}} = (1+v)\overline{\varepsilon_{1,xy}} = \overline{s_{1,xy}} = \overline{s_{1,yx}} \tag{20}$$

Finally, Eqs. (3) and (4) are translated into

$$\bar{\Delta}^2\bar{w}(\bar{r}) - \bar{\varepsilon}_0(1+v)\bar{\Delta}\bar{w}(\bar{r}) - \sum_{x_i,x_j=\bar{x},\bar{y}}\left\{\overline{s_{1,x_ix_j}}(\bar{r})\frac{\partial^2}{\partial x_i\partial x_j}\bar{w}(\bar{r})\right\} = \bar{p} \tag{21}$$

$$\sum_{x_j=\bar{x},\bar{y}}\frac{\partial}{\partial x_j}\overline{s_{1,x_ix_j}}(\bar{r}) = 0, \quad \text{for } x_i = \bar{x}, \bar{y} \tag{22}$$

The boundary conditions as expressed in Eqs. (5) and (6) become

$$\bar{u}(\bar{r}_b) = \bar{v}(\bar{r}_b) = \bar{w}(\bar{r}_b) = 0 \tag{23}$$

and

$$\frac{\partial^2}{\partial x_\perp^2}\bar{w}(\bar{r}_b) = -\bar{K}\frac{\partial}{\partial x_\perp}\bar{w}(\bar{r}_b) \tag{24}$$

The simple conclusion of this lengthy development is that the pressure-dependent deflection profile $\overline{w_{sl}}(\bar{x})$ of an elastically supported rectangular diaphragm made of a single material layer, expressed in terms of dimensionless quantities, takes the form $\overline{w_{sl}}(\bar{x}, \bar{p}) = f_{sl}(\bar{x}, \bar{p}, v, \bar{\varepsilon}_0, \bar{K}, b/a)$, where f_{sl} denotes the solution of the dimensionless equations of equilibrium, i.e. Eqs. (21) and (22) subjected to the boundary conditions Eqs. (23) and (24). In fully dimensional notation, considering that for single layer diaphragms $D_0/D_2 = 12/t^2$, this implies that $w_{sl}(x, p)$ is given by

$$
w_{sl}(x, p) = \sqrt{\frac{D_2}{D_0}} \times f_{sl}\left(\frac{x}{a}, p\frac{a^4}{D_2}\sqrt{\frac{D_0}{D_2}}, v, \varepsilon_0 \frac{a^2 D_0}{D_2}, K\frac{a}{D_2}, \frac{b}{a} \right)
$$

$$
= \frac{t}{\sqrt{12}} f_{sl}\left(\frac{x}{a}, p\frac{a^4}{D_2}\sqrt{\frac{12}{t}}, v, \varepsilon_0 \frac{12a^2}{t^2}, K\frac{a}{D_2}, \frac{b}{a} \right) \tag{25}
$$

In the case of zero applied pressure, i.e. for the analysis of post-buckled structures, inspection of Eqs. (21)–(24) reveals that w_{sl} takes the form

$$
w_{sl}^{(postb.)}(x) = \frac{t}{\sqrt{12}} f_{sl}\left(\frac{x}{a}, 0, v, \frac{12\varepsilon_0 a^2}{t^2}, \frac{Ka}{D_2}, \frac{b}{a} \right) \tag{26}
$$

not explicitly dependent on elastic modulus except through the boundary condition. In most bulge experiments, rigid boundary conditions are a reasonable approximation of reality, corresponding to $K = \infty$. It is concluded that the elastic or plane-strain modulus cannot be extracted from the profiles of post-buckled diaphragms. However, it is in principle possible to determine Poisson's ratio v from the post-buckling profiles of diaphragms with different sizes. A more effective way to achieve this takes advantage of the additional degree of freedom of applying a pressure. This possibility was indeed demonstrated for a PECVD silicon nitride [38]. Similarly, the remaining parameters $D_2 = E_{ps}t^3/12$ and ε_0 can in principle be extracted from load-deflection measurements involving diaphragms of different sizes and aspect ratios.

Note that alternative, yet fundamentally equivalent, renormalization schemes are envisageable. For example, Refs. [6, 36, 37, 75] have built on the following definitions for clamped single-layer diaphragms: $\bar{x}_1 = x_1/a$, $\bar{x}_2 = x_2/a$, $\bar{w} = w/t$ and consequently $\bar{r} = r/a$, $\bar{r}_b = r_b/a$, $\bar{\Delta} = a^2\Delta$, $\bar{u} = ua/t^2$, $\bar{v} = va/t^2$, $\bar{\varepsilon}_i = \varepsilon_i a^2/t^2$ and $\bar{\sigma}_i = \sigma_i a^2/E_{ps}t^2$ with ($i = 0$, xx, xy, yx, yy), and $\bar{p} = pa^4/E_{ps}t^4$. These prescriptions lead to slightly different forms of the out-of-plane equilibrium and boundary conditions, namely

$$
\bar{\Delta}^2 \bar{w}(\bar{r}) - 12\bar{\sigma}_0 \bar{\Delta}\bar{w}(\bar{r}) - 12 \sum_{i,j=1,2}\left[\bar{\sigma}_{1,\bar{x}_i\bar{x}_j}(\bar{r})\frac{\partial^2}{\partial \bar{x}_i \partial \bar{x}_j}\bar{w}(\bar{r}) \right] = \bar{p} \tag{27}
$$

and

$$
\bar{w}(\bar{r}_b) = \frac{\partial}{\partial \bar{x}_\perp}\bar{w}(\bar{r}_b) = 0 \tag{28}
$$

respectively, while all other relationships remain of identical form. The conclusions are identical: the load-deflection response depends on only five independent parameters, namely Poisson's ratio v, the reduced residual strain $\bar{\varepsilon}_0$, the load $\bar{p}$, the rotational suspension stiffness $\bar{K}$, and the geometry via the aspect ratio $b:a$. For a well-defined aspect ratio $b:a$, the four parameters v, $\bar{\varepsilon}_0$, $\bar{p}$ and $\bar{K}$ completely define the load-deflection response. In most cases of interest, the support is well modelled by a rigid clamping condition, with $\bar{K} = \infty$. The load-deflection response of square diaphragms is the special case with $a = b$.

Following the same line of reasoning, clamped, circular single-layer diaphragms are found to follow a load-deflection law of the form $w_{circ}(r) = t \times f_{circ}(r/R, pR^4/tE_{ps}, \varepsilon_0 R^2/t^2, v)$, with the radial coordinate r and the diaphragm radius R.

In the following, when quoting numerical values for reduced quantities resulting from analytical considerations or numerical analysis, we will always indicate the renormalization that leads to the values. With this information, the corresponding value in any other renormalization scheme can be obtained.

3.2.4
Membranes

Very thin diaphragms behave as membranes when the bending term in Eq. (3) is negligible in comparison with the other terms, in particular with the second. This is the case when $\bar{\sigma}_0 = \sigma_0 a^2/E_{ps} t^2$ is sufficiently large, i.e. when at least one of the following four conditions is sufficiently well satisfied: the residual stress σ_0 of the structure is large, its E value is small, its width a is large or its thickness t is small. As an example, with $\bar{\sigma}_0 > 2000$, the response of long membranes at all pressures lies within 2.5% of the exact model developed in Section 3.3.4. In contrast, the membrane approximation fails for weakly tensile and compressive residual stresses and for small and thick diaphragms. The membrane equation derived from Eq. (3) reads

$$t\sigma_0 \Delta w(\mathbf{r}) + t \sum_{i,j=1,2,3} \left[\sigma_{1,ij}(\mathbf{r}) \frac{\partial^2}{\partial x_i \partial x_j} w(\mathbf{r}) \right] = -p \tag{29}$$

It has to be combined with Eqs. (1), (2), (4) and (5). A dimensional analysis shows that the rescaled out-of-plane deflection $\bar{w}(\mathbf{r})/t$ of a rectangular single-layer membrane depends only on the dimensionless parameters b/a, $pa^2/\sigma_0 t^2$, $E_{ps} t^2/\sigma_0 a^2$ and v. The dimensional deflection function of a pressure-loaded membrane therefore takes the form

$$w(\mathbf{r}) = t \times f_{mem}\left(\frac{\mathbf{r}}{a}, \frac{pa^2}{\sigma_0 t^2}, \frac{E_{ps} t^2}{\sigma_0 a^2}, v, \frac{b}{a} \right) \tag{30}$$

Consequently, load-deflection data are analyzed in this case using f_{mem} considering the unknown variables σ_0, E_{ps}, and v as fitting parameters. Since the fitting of v is often numerically ill-defined, it is advisable to assign it a reasonable value and restrict the fitting to the remaining parameters.

3.2.5
General Procedure

In summary, determining the mechanical properties of thin film diaphragms by bulge experiments can proceed as follows:

- Determine the dimensionless solution f_{ps}, f_{sl}, f_{circ}, or other, tailored to the problem.
- Identify the independent parameters, e.g. v, S_0, D_0, D_2, S_1, K and $b:a$, or the relevant subset thereof needed to fully describe the problem. Often K can be assumed to be infinite, corresponding to a clamped diaphragm.
- Perform load-deflection measurements by varying the pressure applied to membranes with different sizes and, possibly, aspect ratios $b:a$. Extract the pressure-dependent load-deflection data and, if necessary, further pieces of information such as the curvature at the edges.
- Extract the values of the independent parameters by using them as fit parameters when fitting the fully dimensional model solutions to the experimental data.

For a single-layer material, E_{ps} and σ_0 will then be obtained from D_0, with $D_2 = D_0 t^2/12$, and from S_0, respectively. If v has been extracted, Young's modulus E follows directly and then so does ε_0 from σ_0.

In the case of multilayer materials, this procedure enables the properties of one material to be obtained if those of the other component layers are known. Use has then to be made of the definitions in Table 3.1, combining the unknown and known properties into the experimentally extracted overall parameters. In order to characterize multilayers fully, therefore, experimental data have to be acquired from diaphragms composed of different subsets of the materials.

Extracting reliable material data from the bulge test therefore requires one to know functions such as f_{ps}, f_{sl}, f_{mem} or f_{circ}. Methods to obtain such functions are described in the next section.

3.3
Load-deflection Models

3.3.1
Introduction

Three methods to obtain load-deflection functions have been used: direct solution of the thin-plate or membrane equations, variational analysis and finite element (FE) analysis.

3.3.1.1 Direct Solution

The first approach consists of explicitly constructing the deformation functions u, v and w. This is possible only in the rarest cases and at the cost of considerable simplifications. Among the explicitly solved problems are circular and rectangular membranes with tensile residual stress at small deflections [5, 9, 20, 78] and long diaphragms under plane-strain deformations [79]. Rectangular tensile membranes in the mechanically linear response regime have been modeled by a series expansion known as the Navier solution [7, 78]. The case of long diaphragms is described in further detail in Section 3.3.4.

3.3.1.2 Variational Analysis

Variational analysis proceeds by constructing approximative solutions by linear superposition of base functions, e.g.

$$u(x, y) = \sum A_{kl} u_{kl}(x, y)$$
$$v(x, y) = \sum B_{mn} v_{mn}(x, y)$$
$$w(x, y) = \sum C_{pq} w_{pq}(x, y) \tag{31}$$

with expansion coefficients A_{kl}, B_{mn} and C_{pq}. These coefficients are obtained by minimizing the total energy U_{total} of the loaded mechanical structure [36].

The total energy

$$U_{total} = U_{strain} + U_{support} - W \tag{32}$$

is composed of terms taking into account the mechanical strain energy U_{strain} of the structure, the deformation energy $U_{support}$ of the support, and the work W performed by the external forces against the plate, which explains the negative sign. In case of rigid clamping, $U_{support} = 0$. In general, the strain energy can be computed as the volume integral

$$U_{strain} = \frac{1}{2} \int_V (\sigma_{xx}\varepsilon_{xx} + \sigma_{yy}\varepsilon_{yy} + 2\sigma_{xy}\varepsilon_{xy}) dV$$

$$= \frac{Et}{2(1+v)} \int_0^t \int_{-b/2}^{b/2} \int_{-a/2}^{a/2} \{\varepsilon_{xx}^2 + \varepsilon_{yy}^2 + 2v\varepsilon_{xx}\varepsilon_{yy} + 2(1-v)\varepsilon_{xy}^2\} dxdydz \tag{33}$$

taking into account that $\sigma_{zz} = \sigma_{xz} = \sigma_{zx} = \sigma_{yz} = \sigma_{zy} = 0$ in view of the definition of a thin plate. For a single-material diaphragm U_{strain} decouples into two terms corresponding to the pure bending and the pure in-plane deformation of the thin structure, denoted U_{bend} and U_{inpl}, respectively. In terms of the deformation functions $u(x,y)$, $v(x,y)$, and $w(x,y)$ of the diaphragm mid-plane, these energy terms are [6, 7, 75, 76]

$$U_{bend} = \frac{D_2}{2} \int\limits_{-b/2}^{b/2} \int\limits_{-a/2}^{a/2} \left\{ \left(\frac{\partial^2 w}{\partial x^2} + \frac{\partial^2 w}{\partial y^2} \right)^2 + 2(1-v) \left[\left(\frac{\partial^2 w}{\partial x \partial y} \right)^2 - \frac{\partial^2 w}{\partial x^2} \frac{\partial^2 w}{\partial y^2} \right] \right\} dxdy \quad (34)$$

and, neglecting constant terms and omitting terms that will evidently integrate to zero in view of the boundary conditions,

$$U_{inpl} = \frac{D_0}{2} \int\limits_{-b/2}^{b/2} \int\limits_{-a/2}^{a/2} \left\{ \left(\frac{\partial u}{\partial x} \right)^2 + \left(\frac{\partial v}{\partial y} \right)^2 + \frac{1}{2}(1-v) \left[\left(\frac{\partial u}{\partial y} \right)^2 + \left(\frac{\partial v}{\partial x} \right)^2 \right] \right.$$

$$+ 2v \frac{\partial u \partial v}{\partial x \partial y} + (1-v) \frac{\partial u \partial v}{\partial y \partial x} + \frac{\partial u}{\partial x} \left(\frac{\partial w}{\partial x} \right)^2 + \frac{\partial v}{\partial y} \left(\frac{\partial w}{\partial y} \right)^2$$

$$+ v \left[\frac{\partial u}{\partial x} \left(\frac{\partial w}{\partial y} \right)^2 + \frac{\partial v}{\partial y} \left(\frac{\partial w}{\partial x} \right)^2 \right] + (1-v) \left(\frac{\partial u}{\partial y} + \frac{\partial v}{\partial x} \right) \frac{\partial w}{\partial x} \frac{\partial w}{\partial y}$$

$$+ (1+v)\varepsilon_0 \left[\left(\frac{\partial w}{\partial x} \right)^2 + \left(\frac{\partial w}{\partial y} \right)^2 \right] + \frac{1}{4} \left(\frac{\partial w}{\partial x} \right)^4 + \frac{1}{4} \left(\frac{\partial w}{\partial y} \right)^4$$

$$\left. + \frac{1}{2} \left(\frac{\partial w}{\partial x} \right)^2 \left(\frac{\partial w}{\partial y} \right)^2 \right\} dxdy \quad (35)$$

The integrals extend over the in-plane dimensions of the diaphram from $-a/2$ to $a/2$ and from $-b/2$ to $b/2$. Similarly, W is

$$W = p \int\limits_{-b/2}^{b/2} \int\limits_{-a/2}^{a/2} w(x,y)dxdy \quad (36)$$

which is the area integral of force, i.e. $pdxdy$, times distance, w. It is seen that U_{inpl} contains terms involving ε_0 and the in-plane and out-of-plane displacements u, v and w. In contrast, U_{bend} and W depend only on the out-of-plane displacement w. In the case of thin diaphragms, U_{bend} is negligible in comparison with the other terms and can be omitted from the energy minimization.

The energy minimization procedure is illustrated with the simplest case where the variational space is restricted to a single base function per displacement direction, i.e. $u(x,y) = Au_{var}(x,y)$, $v(x,y) = Bv_{var}(x,y)$ and $w(x,y) = Cw_{var}(x,y)$. When these are inserted into U_{total}, it is seen that U_{bend} contributes terms proportional to C^2. Analogously, U_{inpl} is composed of terms proportional to A, A^2, B, B^2, AC^2, BC^2 and C^4. Finally, W is proportional to C. The total energy therefore has the structure

$$U_{total} = \frac{D_0}{2}(I_{A2}A^2 + I_{B2}B^2 + I_{AB}AB + I_{AC2}AC^2 + I_{BC2}BC^2 + I_{C4}C^4)$$

$$+ \left[\frac{D_2}{2} I_{C2a} + \frac{D_0}{2}(1+v)\varepsilon_0 I_{C2b} \right] C^2 - pI_C C \quad (37)$$

where the coefficients I_0, I_A, I_{A2}, etc., are calculated according to the integrals in Eqs. (34)–(36). The minimization procedure requires U_{total} to be varied independently with respect to all three parameters A, B and C, and the global minimum to be found. This is started by taking the derivatives of U_{total} with respect to A and B and equating them to zero. This provides conditions relating A and B to ε_0 and C, namely $A = C^2(2I_{AC2}I_{B2} - I_{AB}I_{BC2})/(I_{AB}^2 - 4I_{A2}I_{B2})$ and $B = C^2(2I_{BC2}I_{A2} - I_{AB}I_{AC2})/(I_{AB}^2 - 4I_{A2}I_{B2})$. These conditions show that the in-plane displacements vary quadratically with the out-of-plane deflection. In turn, this implies, for instance, that upward and downward post-buckling are energetically equivalent. When these conditions are inserted into Eq. (37), U_{total} takes the form

$$U_{total} = \frac{D_0}{2}\left(I_{C4} + \frac{I_{AC2}^2 I_{B2} + I_{BC2}^2 I_{A2} - I_{AB}I_{AC2}I_{BC2}}{I_{AB}^2 - 4I_{A2}I_{B2}}\right)C^4$$
$$+ \left\{\frac{D_2}{2}I_{C2a} + \frac{D_0}{2}(1+v)\varepsilon_0 I_{C2b}\right\}C^2 - pI_C C \qquad (38)$$

In the last step of the energy minimization procedure, Eq. (38) is varied with respect to C, which means that its derivative is set equal to 0. The result is the load-deflection law

$$p = \left(D_2 \frac{I_{C2a}}{I_C} + \sigma_0 t \frac{I_{C2b}}{I_C}\right)C + 2D_0\left[\frac{I_{C4}}{I_C} + \frac{I_{AC2}^2 I_{B2} + I_{BC2}^2 I_{A2} - I_{AB}I_{AC2}I_{BC2}}{I_C(I_{AB}^2 - 4I_{A2}I_{B2})}\right]C^3 \qquad (39)$$

where use was made of the connection between initial strain and stress of the diaphragm material, i.e. $\sigma_0 = E\varepsilon_0/(1-v)$.

Many of the general properties of the non-linear deflection of clamped thin diaphragms are contained in Eq. (39):

- At small pressures, the deflection proportional to C depends linearly on the applied load; the resistance against the deflection results from the bending stiffness and from the prestress, as expressed by the term linear in C.
- At larger pressures, the non-linear terms increasingly impede the further deflection of the diaphragm.
- For negative initial stresses, the linear term is negative and non-trivial solutions exist even in the absence of a load; these are the post-buckling deflections; the critical buckling prestress is deduced from the assumed variational functions is $\sigma_{0,cr} = -D_2 I C_{2a}/I C_{2b}t$.

In the case of more encompassing expansion series defining the variational space of the functions u, v and w, the individual terms in Eq. (37) are replaced by products of the coefficients A_{kl}, B_{mn} and C_{pq} from the corresponding displacement functions. The coefficients A_{kl} and B_{mn} are found to be sums of second-order products of the C_{pq} coefficients. In the analogy of Eq. (38), U_{total} consists of sums of second-order and fourth-order terms in the C_{pq} coefficients from the strain energy and terms linear in C_{pq} and proportional to p, from W. Finally, the variation with respect to all the C_{pq} coefficients would result in a system of coupled nonlinear equations for these coefficient, depending on the pressure. This system of equations could then in principle be solved by numerical methods. Instead, the minimization of U_{total} with respect to the coefficients C_{pq} is usually carried out directly.

For membranes, trial functions such as

$$u_{kl}(x, y) = A_{kl}\sin(k2\pi x/a)\cos(l\pi y/b), \text{ with } k = 1, 2, 3, \ldots \text{ and } l = 1, 3, 5, \ldots$$
$$v_{mn}(x, y) = B_{mn}\cos(m\pi x/a)\sin(n2\pi y/b), \text{ with } m = 1, 3, 5, \ldots \text{ and } n = 1, 2, 3, \ldots$$
$$w_{pq}(x, y) = C_{pq}\cos(p\pi x/a)\cos(q\pi y/b), \text{ with } p = 1, 3, 5, \ldots \text{ and } q = 1, 3, 5, \ldots$$

$$(40)$$

have served the purpose sufficiently well. In the early bulge test reports, only the terms with $k = l = m = n = p = q = 1$ of this system were retained [11, 15, 56]. For membranes with large aspect ratio $b:a$, this early truncation leads to considerable errors, since the plane-strain response in their middle section with steep descents towards the smaller diaphragm edges is unsatisfactorily rendered. For both x and y directions, we recommend including functions with wavelengths down to at least $a/5$, where a denotes the smaller of the two side lengths.

The test functions in Eq. (40) do not satisfy clamping boundary conditions. For clamped thin plates, it is better to design test functions each satisfying the required boundary conditions. For this purpose, alternative test functions were implemented and were shown to provide accurate results over a wide range of experimental conditions [6, 23, 36–38]. These functions were tailored to the symmetries of the observed deflection shapes. Up to 2×16^2 trial functions for the in-plane deformations and 8^2 functions for the out-of-plane deformation were included in the most advanced case [36].

Implementing the corresponding total energy and minimizing it with respect to large numbers of coefficients of the linear combinations of test functions represents a formidable task. Careful book-keeping of millions of energy terms is essential [36, 75]. For the minimization, the method of conjugated gradients was found to be efficient [80].

3.3.1.3 Finite-element Analysis

The third approach is finite-element analysis. After the diaphragm structure has been meshed, an approximation to the deformation functions u, v and w is obtained by computing the three-dimensional deflections of the mesh nodes. The challenge is to use well-suited finite elements and to define a mesh adequately mapping the geometric situation. If shell elements are used, the FE software has to be able to handle the geometric nonlinearity represented by the strain-deformation relations in Eq. (1). Further, it has to be able to deal with bifurcations such as buckling and other symmetry transitions. Convergence is often found to be a delicate issue and has to be considered carefully. The software packages ANSYS [34, 36, 39] and ABAQUS and ADINA [9] have been employed for bulge test simulations.

3.3.2
Square Diaphragms

The load-deflection behavior of square diaphragms is illustrated in Figure 3.2. The dimensionless central deflection amplitude $\overline{w}_0$ is shown schematically as a function of reduced residual strain $\overline{\varepsilon}_0$ and load pressure $\overline{p}$. Optical micrographs of membranes illustrate the various types of deflection profiles. Vertical dashed lines, e.g. from (c) in Figure 3.2 via (f) or from (d) via (g), (h) and (i) correspond to load-deflection experiments, i.e. they show the evolution of $\overline{w}_0$ of a diaphragm with fixed $\overline{\varepsilon}_0$ as a function of $\overline{p}$. Curves of constant pressure values, e.g. $\overline{p}_1$, $\overline{p}_2$, ..., describe the deflection of diaphragms with different reduced residual strains under identical reduced loads. Diaphragms with $\overline{\varepsilon}_0 > \overline{\varepsilon}_{0,cr1}$, i.e. weakly compressive or tensile residual strain, are flat in the unloaded state (Figure 3.2a). A pressure load progressively bulges them out of the plane. The isobaric line $\overline{p} = 0$ to the left of $\overline{\varepsilon}_{0,cr1}$ shows the post-buckling deflection of the structures [36]. Their unloaded equilibrium position is in a bulged state (Figure 3.2a–e), i.e. $\overline{w}_0 \neq 0$ at $\overline{p} = 0$. Their energy is lowered by a partial out-of-plane relaxation of the in-plane stress, at the

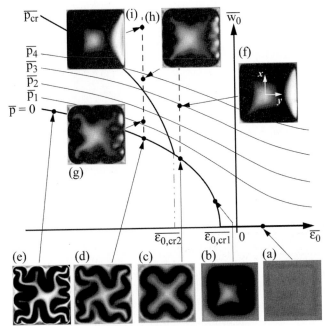

Figure 3.2 Schematic diagram of the load-deflection response of square clamped diaphragms, i.e. reduced dimensionless center deflection $\overline{w}_0$ as a function of reduced residual strain $\overline{\varepsilon}_0$ and pressure $\overline{p}$, illustrated by optical micrographs of silicon nitride diaphragm deflection profiles. Adapted from [75].

cost of a slight increase in bending energy. The value of $\varepsilon_{0,\text{cr1}}$ has been obtained as $\varepsilon_{0,\text{cr1}} \cong -4.363t^2/a^2(1 + v)$, corresponding to the critical residual stress $\sigma_{0,\text{cr1}} = -4.363E_{\text{ps}}t^2/a^2$ [36].

At pressures higher than the $\overline{\varepsilon}_0$-dependent critical pressure $\overline{p_{\text{cr}}}(\overline{\varepsilon}_0)$ and above the isobar $\overline{p} = 0$, the deflection profiles show all the in-plane reflection and rotation symmetries of the square. At pressures below $\overline{p_{\text{cr}}}$, the profiles show only the in-plane symmetry under rotations by multiples of $\pi/2$ (Figure 3.2d, e). The pressure $\overline{p_{\text{cr}}}$ thus corresponds to a symmetry transition analogous to the buckling transition at $\overline{\varepsilon}_{0,\text{cr1}}$ for unloaded diaphragms [39]. The critical pressure line $\overline{p_{\text{cr}}}(\overline{\varepsilon}_0)$ intercepts the post-buckling line $\overline{p} = 0$ at the critical reduced residual strain $\overline{\varepsilon}_{0,\text{cr2}}$. Thus, to the left of $\overline{\varepsilon}_{0,\text{cr2}}$ unloaded post-buckled profiles show the reduced symmetry, whereas to the right they show the full set of symmetries of a square. Values of $\varepsilon_{0,\text{cr2}}a^2/t^2$ depend on v and were found to be, e.g., -212 for $v = 0.125$, -206 for $v = 0.25$ and -208 for $v = 0.375$ [36]. Values of $\overline{p_{\text{cr}}}(\overline{\varepsilon}_0)$ away from the $\overline{p} = 0$ line have proved hard to determine reliably due to convergence problems with the FE method and excessive numbers of test functions necessary to model these structures accurately using the variational method [39].

3.3.2.1 Approximate Load-deflection Laws

The mechanical response of square structures to pressures has been extensively analyzed using the FE method. Based on the numerical results, the following load-deflection law was found to fit the numerical results best [6]:

$$p(w_0) = E_{\text{ps}}(1 + v)\left[c_1(\varepsilon_0 - \varepsilon_{0,\text{cr1}})\frac{tw_0}{a^2} + c_3(v)\left(1 + d_3\varepsilon_0\frac{a^2}{t^2}\right)\frac{tw_0^3}{a^4}\right] \tag{41}$$

with $c_1 = 13.84$, $c_3(v) = 32.852 - 9.48v$ and $d_3 = -2.6 \times 10^{-5}$. This result was extracted from simulations covering the parameter space $0.15 < v < 0.35$ and $-200 < \varepsilon_0a^2/t^2 < 1000$. If ε_0 is extracted from load-deflection curves, its value lies within a distance of less than $5(t/a)^2$ from the true ε_0. Further, relative uncertainties of the extracted E value are smaller than 1% over the entire parameter range. For circumstances where σ_0 rather than ε_0 is to be extracted, Eq. (41) is equivalent to

$$p(w_0) = \frac{c_1}{a^2}\left(t\sigma_0 + 4.363E_{\text{ps}}\frac{t^3}{a^2}\right)w_0 + \frac{c_3(v)}{a^4}(1 + v)\left(tE_{\text{ps}} + d_3\sigma_0\frac{a^2}{t}\right)w_0^3 \tag{42}$$

As a consequence of Eq. (25), the residual strain of a compressive diaphragm material can be extracted from the central deflection of a square alone, if Poisson's ratio of the material is known. For residual strains between $\varepsilon_{0,\text{cr1}}$ and $\varepsilon_{0,\text{cr2}}$, Ref. [38] describes a suitable method to reach this goal. Using the variational method, the relationship between reduced central deflection and reduced residual strain was found to be well approximated by

$$w_0 = a\Delta\varepsilon_0^{1/2}\left[c_1 + c_2\tanh\left(c_3\Delta\varepsilon_0\frac{a^2}{t^2}\right) + \frac{c_4\Delta\varepsilon_0\dfrac{a^2}{t^2} + c_5\Delta\varepsilon_0^2\dfrac{a^4}{t^4}}{1 - c_6\Delta\varepsilon_0^3\dfrac{a^6}{t^6}}\right]^{1/2} \tag{43}$$

with $\Delta\varepsilon_0 = \varepsilon_{0,\mathrm{cr1}}(V) - \varepsilon_0$ and

$$c_1 = -0.4972 - 0.2313v - 0.2128v^2$$
$$c_2 = 0.0698 + 0.1625v + 0.2v^2$$
$$c_3 = -7.19 \times 10^{-3} - 0.0466v + 0.0367v^2$$
$$c_4 = -1.19 \times 10^{-3} + 5.51 \times 10^{-3}v^2$$
$$c_5 = -3.34 \times 10^{-6} - 7.43 \times 10^{-5}v + 1.28 \times 10^{-4}v^2$$
$$c_6 = 3.16 \times 10^{-6} + 4.8 \times 10^{-6}v - 1.52 \times 10^{-5}v^2$$

3.3.2.2 Square Membranes

Square diaphragms under strongly tensile residual stresses or strong loads behave as membranes. These were among the first MEMS structures analyzed using the bulge test [10, 11, 15, 19]. For this purpose, the variational functions $u(x,y) = A\sin(2\pi x/a)\cos(\pi y/a)$, $v(x,y) = B\cos(\pi x/a)\sin(2\pi y/a)$ and $w(x,y) = C\cos(\pi x/a)\cos(\pi y/a)$ were inserted into U_{total}, i.e. Eq. (32). The corresponding approximate load-deflection law is of the form

$$p(w_0) = c_1\frac{\sigma_0 t}{a^2}w_0 + c_3(v)\frac{Et}{(1-v)a^4}w_0^3 \tag{44}$$

The two numerical coefficients c_1 and $c_3(v)$ have been determined in several studies. Values are summarized in Table 3.2. Residual stresses and strains and elastic coefficients of various materials extracted using the bulge test of square membranes are listed in Tables 3.5–3.9 in Section 3.5.

3.3.3
Rectangular Diaphragms

Rectangular diaphrams were analyzed in the membrane limit. Variational calculations based on the test functions $u(x,y) = A\sin(2\pi x/a)\cos(\pi y/b)$, $v(x,y) = B\cos(\pi x/a)\sin(2\pi y/b)$ and $w(x,y) = C\cos(\pi x/a)\cos(\pi y/b)$ yielded a load-deflection law of structure identical with that of Eq. (44) with coefficients c_1 and c_3 given by [15]

$$c_1(n, v) = \frac{\pi^4(1+n^2)}{16} \tag{45}$$

and

$$c_3(n, v)$$
$$= \frac{\pi^6}{2(1-v^2)}\left(\frac{9+2n^2+9n^4}{256} - \frac{[4+n+n^2-4n^3-3nv(1+n)]^2}{2\{81\pi^2(1+n^2)+128n+v[128n-9\pi^2(1+n^2)]\}}\right) \tag{46}$$

Table 3.2 Coefficients c_1 and c_3 in Eq. (44), as reported in the bulge test literature. In the quoted sources, the membrane width is usually defined as $2a$. This explains the apparent discrepancy by a factor of 4 in c_1 and 16 in c_2 between published and tabulated values.

Geometry	c_1	c_3	Type of analysis, comment [source]
Square	12.1761	21.9405	Variational, $v = 0.25$ [15]
	$\pi^4/8$	$\dfrac{\pi^6}{2(1+v)}\left[\dfrac{5}{64} + \dfrac{(5-3v)^2}{9\pi^2(v-9)-64(1+v)}\right]$	Variational [15]
	13.64	$31.7 - 9.36v$	Finite elements [9]
	13.572	$16/(0.8 + 0.062v)^3$	Variational [21]
	13.8	$31.9 - 8.65v$	Variational [24]
Rectangular	8	$64/3(1 + v)$	FE [13], variational [21], analytical
	–	$21.6(1.41 - 0.292v)$	Clamped [43]
		See Section 3.3.3	Variational [15]
Circular	16	$2^7/3$	Insert diameter $2r$ instead of length a in Eq. 44[5]

where $n = a/b$ and b denotes the longer side length of the membranes. For $n = 1$, these results coincide with those cited in the first two lines in Table 3.2. In the limit case of long membranes ($n = 0$), the asymptotic expressions are $c_1(0,v) = \pi^4/16$ and $c_3(0,v) \cong (108.802 - 16.8994v)/[(9 - v)(1 + v)]$. These values deviate strongly from the results obtained analytically for long membranes, as described in the next section. The discrepancy results from the inadequacy of the used test functions in modelling the plane-strain deflection shape of long membranes.

It is therefore recommended to apply the c_1 and c_3 expressions in Eqs. (45) and (46) only for membranes that are slightly rectangular, up to an aspect ratio of maybe $b:a = 2:1$. For membranes and thin plates with aspect ratios $b:a$ larger than 6:1, the theory of long membranes developed in the next section is definitely more appropriate for the sections deformed in plane strain.

3.3.4
Long Diaphragms with Rigid Supports

An extremely useful geometry for bulge testing is provided by long diaphragms. These are structures with large aspect ratios $b:a$. The small edges negligibly influence the mechanical response of an extended central section of the structure. There, the load-deflection profile is invariant under continuous or discrete translations in the longitudinal direction. This part of the structure is appropriately modelled by an infinitely long diaphragm.

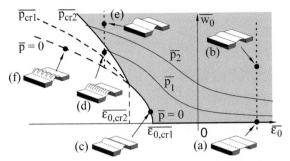

Figure 3.3 Schematic load-deflection response diagram of long clamped diaphragms, i.e., reduced dimensionless deflection amplitude $\overline{w}_0$ as a function of reduced residual strain $\overline{\varepsilon}_0$ and pressure $\overline{p}$, illustrated by selected post-buckling and deflection profiles.

The mechanical response of such structures is shown schematically in Figure 3.3. As in Figure 3.2, the reduced center deflection amplitude $\overline{w}_0$ is shown as a function of the reduced residual stress $\overline{\varepsilon}_0$ and pressure $\overline{p}$. Selected states are illustrated by schematic three-dimensional deflection profiles. Vertical lines correspond to load-deflection experiments, showing the progressive increase of $\overline{w}_0$ with increasing pressure, e.g. from (a) to (b) or from (d) to (e).

Structures with tensile or weakly compressive residual stress ($\overline{\varepsilon}_0 > \overline{\varepsilon}_{0,cr1}$) respond to external differential pressures by a translationally invariant plane-strain deformation. Without load, the diaphragm is flat (Figure 3.3a). Under pressure it progressively bulges out of the plane (Figure 3.3c). At the critical reduced residual strain value $\varepsilon_{0,cr1} = -\pi^2 t^2/3a^2(1 + \nu)$ corresponding to the critical residual stress $\sigma_{0,cr1} = -\pi^2 E_{ps} t^2/3a^2$, the structures undergo a buckling transition, below which they are post-buckled even for $p = 0$. For ε_0 values below the second critical residual strain value $\varepsilon_{0,cr2} \approx -17.3 t^2/a^2$, the unloaded structures show a meander-like superstructure superimposed on their plane-strain profile. Under pressure the meanders are progressively modified until they cross over into ripple-shaped profiles at a critical pressure level $\overline{p}_{cr1}(\overline{\varepsilon}_0, \nu)$ [37, 40]. With increasing pressure, the ripple amplitude decreases and finally vanishes completely at a second critical pressure $\overline{p}_{cr2}(\overline{\varepsilon}_0, \nu)$. This second transition offers the possibility to determine Poisson's ratio of the material [38]. An example of such a profile evolution is shown in Figure 3.4.

3.3.4.1 Plane-strain Response

In the plane-strain regime, i.e. for $\overline{p} > \overline{p}_{cr2}(\overline{\varepsilon}_0, \nu)$, Eqs. (3) and (4) simplify to a one-dimensional problem [31], as described in Section 3.2.1 with Eqs. (10) and (11). These equations have to be solved for $w(x)$ subject to the boundary conditions, i.e. Eqs. (5) and (6). For many practical purposes, the diaphragm support can be assumed as stiff, so that $dw/dx = 0$ at the boundary. The deflection profile solving this problem is

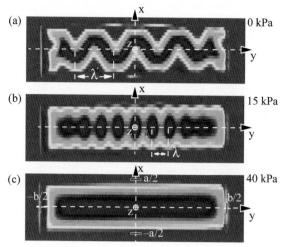

Figure 3.4 Measured profiles of a 179 μm wide, 697 μm long and 490 nm thick PECVD silicon nitride membrane at three different pressure loads: (a) meander profile without load; (b) ripple profile at 15 kPa; (c) plane-strain profile at 40 kPa [37].

$$w(x) = \frac{pa^2}{2S}\left[\frac{1}{4} - \left(\frac{x}{a}\right)^2 - \frac{\cosh\left(a\sqrt{\frac{S}{4D}}\right) - \cosh\left(x\sqrt{\frac{S}{D}}\right)}{a\sqrt{\frac{S}{D}}\sinh\left(a\sqrt{\frac{S}{4D}}\right)} \right] \tag{47}$$

while the load-deflection law resulting from this solution is

$$p = \frac{8}{\frac{a^2}{S} - 4a\sqrt{\frac{D}{S^3}}\tanh\left(\frac{a}{4}\sqrt{\frac{S}{D}}\right)} w_0 \tag{48}$$

where the line stress S solves

$$S = \sigma_0 t + \frac{24D}{a^2} \times$$

$$\frac{\left[\left(8 + a^2\frac{S}{3D}\right)\sinh\left(a\sqrt{\frac{S}{4D}}\right)^2 - a^2\frac{S}{2D} - 3a\sqrt{\frac{S}{4D}}\sinh\left(2a\sqrt{\frac{S}{4D}}\right)\right]}{\left[a\sqrt{\frac{S}{4D}}\sinh\left(a\sqrt{\frac{S}{4D}}\right) - 4\sinh\left(a\sqrt{\frac{S}{4D}}/2\right)^2\right]^2} \frac{w_0^2}{2} \tag{49}$$

namely Eq. (11) in explicit form. Equations (48) and (49) represent the exact nonlinear load-deflection law $p(w_0, \sigma_0, E_{ps}, a, t)$ of long, thin plates with clamped edges under plane-strain deformation. One way of obtaining E_{ps} and σ_0 from a set

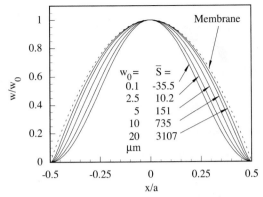

Figure 3.5 Deflection profiles calculated using Eqs. (47), (48), and (49) for various effective stress values $\bar{S} = \sigma t a^2 / D_2$, showing the transition from plate to membrane behavior (adapted from [31]).

of experimental data $\{p_i, w_{0,i}\}$ with $i = 1, \ldots, N$, is thus to fit the load-deflection law to the data, with σ_0 and E_{ps} used as fitting parameters. The evolution of deflection profiles as a function of applied pressure is illustrated in Figure 3.5. With increasing effective stress parameter $\bar{S} = \sigma t a^2 / D_2$, the profile undergoes a transition from plate to membrane behavior.

On the buckling branch, for reduced residual strains $\bar{\varepsilon}_0$ between $\bar{\varepsilon}_{0,cr2}$ and $\bar{\varepsilon}_{0,cr1}$, the center deflection grows as

$$|w_0| = \frac{2t}{\sqrt{3}} \left(\frac{\sigma_0}{\sigma_{0,cr1}} - 1 \right)^{1/2} = \frac{2t}{\sqrt{3}} \left(\frac{\varepsilon_0}{\varepsilon_{0,cr1}} - 1 \right)^{1/2} \tag{50}$$

and the deflection profile is $w(x) = w_0[1 + \cos(2\pi x/a)]/2$ [31]. On this entire post-buckling branch ($\bar{p} = 0$), the mid-plane stress is $\sigma_{0,cr1}$.

3.3.4.2 Long Membranes

With a strongly tensile residual stress, the structure shows membrane behavior. Its deflection profile is

$$w(x) = \frac{p}{2t\sigma} \left[\left(\frac{a}{2} \right)^2 - x^2 \right] \tag{51}$$

where $\sigma = \sigma_0 + 8 E_{ps} w_0^2 / 3a^2$ is the effective stress acting on the mid-plane of the deflected diaphragm. With this, the load-deflection law of long membranes [31, 75]

$$p(w_0) = \frac{8\sigma_0 t}{a^2} w_0 + \frac{64 E_{ps} t}{3 a^4} w_0^3 \tag{52}$$

is obtained. Residual stress σ_0 and plane-strain modulus E_{ps} can thus be extracted from the linear and cubic components of the membrane response.

In situations with large effective stress, the load-deflection law implicit in Eqs. (48) and (49) merges into the membrane response of Eq. (52). This is the case with films subjected to strongly tensile stress. However, it is also true for membranes under weakly tensile and even compressive residual stress at sufficiently high pressure loads. The effective stress σ has then evolved to a highly tensile level and the membrane approximation is comfortably used.

The meanders and ripples in long diaphragms under strongly compressive residual stress and the corresponding hierarchy of symmetry transitions under pressure were analyzed using the energy minimization method [37, 40]. The out-of-plane profile was decomposed into plane-strain, undulating and skew-symmetric components. The superposition of the three types of functions produces the observed meander-shaped pattern. With increasing pressure, the coefficients of the skew-symmetric components progressively decrease. At p_{cr1} they vanish and remain equal to zero beyond. Thus the remaining profile is composed of plane-strain and undulating components combining into ripple-shaped profiles, as observed. With further pressure increase, the coefficients of the undulating component progressively shrink and finally vanish at p_{cr2}. Beyond this point, the deflection profile reduces to the plane-strain components. Experimental profiles of a diaphragm at these three stages are shown in Figure 3.4.

3.3.4.3 Extraction of Poisson's Ratio

The last symmetry transition, i.e. from ripples to plane strain, has been analyzed extensively using the energy minimization method. It was shown to depend strongly on Poisson's ratio v. Qualitatively, the reason is simple: the ripple manifests the tendency of the diaphragm to relax the longitudinal residual stress component; under pressure, the additional transverse strain is translated via Poisson's ratio v into a longitudinal contraction, which ultimately compensates the longitudinal residual stress component and flattens out the structure in the longitudinal direction. Beyond, the diaphragm is left in a plane-strain state. Evidently, the dependence of this transition depends on v: for $v = 0.5$, the longitudinal contraction is strongest and the pressure at which compensation occurs is lowest; at the other extreme, for $v = 0$ orthogonal in-plane strains are decoupled and no pressure is able to straighten the profile. The problem was quantitatively addressed by an instability analysis of plane-strain profiles with respect to the addition of undulating components [38]. The analysis showed that the dependence of p_{cr2} and corresponding deflection $w_{0,cr2}$ on v is sufficiently strong to enable v of thin films to be stably determined. The ripple wavelength constitutes a further parameter serving as an independent numerical self-consistency check. A typical experiment proceeds by first determining the transition pressure and deflection, i.e. p_{cr2} and $w_{0,cr2}$, respectively, then by measuring the load-deflection response of the plane-strained diaphragm at pressures $p > p_{cr2}$ resulting in σ_0 and E_{ps}, and finally extracting Poisson's ratio. Using this method, v of a PECVD silicon nitride film with σ_0 between -57.8 and -82.3 MPa was determined as 0.253 ± 0.017 [38].

An alternative method to determine v combines bulge experiments on square and long diaphragms. The different load-deflection laws of these two geometries makes it possible to determine different combinations of elastic coefficients, from which v is finally extracted. This was achieved with an unstressed PECVD silicon nitride with $\sigma_0 = -(2.4 \pm 2.4)$ MPa where $v = 0.235 \pm 0.04$ was found [6]. Previous attempts aiming in the same direction have possibly suffered from the lack of sufficiently accurate model for the load-deflection response of square and long diaphragms, including their bending rigidity [81].

A final comment concerns the question of how long is long enough. For the evaluation of parameters from profiles under plane strain, diaphragm aspect ratios $b:a$ larger than $6:1$ provide reliable results; for the extraction of parameters from rippled and meandering profile, aspect ratios larger than 12 are recommended. If the structure is too short, wavelength selection disturbs the profile and influences the extracted properties. Residual stresses and strains and elastic coefficients of various materials extracted using the bulge test of long membranes are given in Section 3.5.

3.3.5
Long Single-layer Diaphragms with Elastic Supports Under Plane Strain

In the case of long membranes with compliant supports, the appropriate boundary conditions are those given by Eqs. (23) and (24). The normalized load-deflection relation is then [31]

$$\bar{w}(\bar{x}) = \frac{\bar{p}}{2\bar{S}}\left[\frac{1}{4} - \bar{x}^2 - (2 + \bar{K})\frac{\cosh(\sqrt{\bar{S}}/2) - \cosh(\sqrt{\bar{S}}\bar{x})}{\bar{S}\cosh(\sqrt{\bar{S}}/2) + \bar{K}\sqrt{\bar{S}}\sinh(\sqrt{\bar{S}}/2)}\right] \tag{53}$$

The normalization scheme applied in this section is the first in Section 3.2.3. Note that Eq. (53) can also be used when dealing with multilayer diaphragms as long as the absence of initial bending moments is insured ($\bar{S}_1 = 0$). From the definition of $\bar{S}_1$ (see Table 3.1), this condition is always met by structures composed of a single material. An obvious conclusion that can be drawn from Eq. (53) is that the deflection response is symmetric around $\bar{p} = 0$, which is not necessarily true for multilayers, as demonstrated in the next section. In the limit of large values of $\bar{K}$, and as expected, the current solution tends to that given by Eq. (47), whereas for sufficiently small values of $\bar{K}$, the membrane is mechanically equivalent to a simply supported structure. Equation (53) also provides the information necessary to grasp the range of values of $\bar{K}$ that continuously transform a structure with compliant supports to one with both ends rigidly fixed: the buckling instability occurs at those negative values of $\bar{S}$ where the function $\bar{S}\cosh(\sqrt{\bar{S}}/2) + \bar{K}\sqrt{\bar{S}}\sinh(\sqrt{\bar{S}}/2)$ in the denominator approaches zero, that is, when

$$\bar{S} = \bar{S}_{\text{inst}} = -4[\tan c^{-1}(2/\bar{K})]^2 \tag{54}$$

Here $\tan c^{-1}$ denotes the inverse function of the cardinal tangent, i.e. $\tan c(x) = \tan(x)/x$ for $x \neq 0$ and $\tan c(0) = 1$. The predicted variation of $\bar{S}_{inst}$ with $\bar{K}$ is shown in Figure 3.6. For $\bar{K} < 10^{-2}$, it is clear that the buckling line force is constant and equal to $-\pi^2$. This corresponds to a structure with compliant supports. At the other extreme, $\bar{S} = -4\pi^2$ for values of $\bar{K}$ larger than 10^4, i.e. for a fully clamped diaphragm. Between these values, a continuous variation occurs depending on the membrane size and thickness and plane-strain modulus of the component layers $[\bar{K} = Ka/D_2 = E_{ps}^{(substr.)}at^2/D_2k_r]$. The bulge technique is commonly applied to diaphragms with widths a in the range of hundreds of μm and thicknesses t running from a few tenths of a μm to several μm [1, 5, 22, 31, 82]. For this matter, most of the inorganic devices fall in the fully clamped case and little boundary softening can be observed with such structures. For illustration, the normalized stiffness of the supporting edges $\bar{K}$ of a $1\,\mu m$ thick polycrystalline silicon membrane ($E_{ps} \cong 160\,GPa$ [46, 82]) will decrease from 10^4 to only 10^3 by decreasing its width from $1\,mm$ to $100\,\mu m$. Nevertheless, it is worthwhile mentioning that this variation leads to a more accurate description of the membrane mechanics with corrections on the predicted stress distributions of a few percent, which in turn are very important for fracture analysis, for instance, as explained in Section 3.3.7. The support stiffness parameter $\bar{K}$ scales linearly with the width-to-thickness ratio of the diaphragms and is inversely proportional to Young's modulus of the structural films. As a consequence, polymer-based devices can always be considered as structures with rigid clamping supports, due to their lower plane-strain modulus in comparison with inorganic films.

In order to illustrate in more detail the influence of compliant supports in the mechanical response of membranes under applied differential pressure, normalized isobaric curves of the center deflection $\bar{w}_0 = \bar{w}(\bar{x} = 0)$ as a function of the

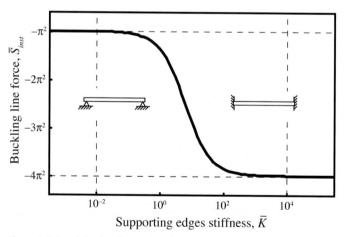

Figure 3.6 Instability line force plotted as a function of the stiffness of the membrane supporting edges. Simply supported diaphragms are obtained when $\bar{K} < 10^{-2}$, whereas $\bar{K} > 10^4$ defines rigidly clamped structures.

prestress, for different clamping conditions, are shown in Figure 3.7. The first case, Figure 3.7a, corresponds to a fully clamped structure (large $\bar{K}$) with no initial bending moments ($\bar{S}_1 = 0$). The bifurcation in the ($\bar{S}_0$, $\bar{w}_0$) diagram occurring at $(-4\pi^2, 0)$ indicates that, at no applied external force, a membrane will rest in a flat position if its reduced initial line force is larger than $-4\pi^2$, whereas for values below $-4\pi^2$, a split in the response is observed and the membrane undergoes Euler-like post-buckling [31]. In the second graph, Figure 3.7b, a simply supported case is shown still for a structure with $\bar{S}_1 = 0$. Qualitatively, Figures 3.7a and b show similar dependences of the center deflection with prestress. The main differences

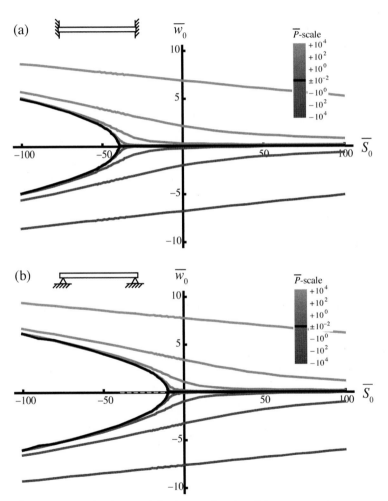

Figure 3.7 Normalized center deflection as a function of the reduced initial line force at different pressures for structures with no initial bending moments ($\bar{S}_1 = 0$) and with (a) rigid supports ($\bar{K} = 10^4$) and (b) compliant clamps ($\bar{K} = 10^{-2}$).

reside in the prestress position of the buckling bifurcation, which is shifted from $-4\pi^2$ to $-\pi^2$, in agreement with Eq. (54) and Figure 3.6, when the supports become more compliant.

3.3.6
Long Multilayer Diaphragms with Elastic Supports Under Plane Strain

This section describes the mechanics of multilayer membranes with compliant supports. The analysis of multilayer structures is motivated by the fact that devices are usually composed of stacks of layers of different films rather than of a single thin film. Further, the bulge characterization can be difficult when dealing with membranes with significant levels of compressive residual stress since those structures often present complex buckled shapes [6, 38–40, 75]. Nevertheless, by combining such films with tensile layers, stress compensation is achieved and the analysis is simplified.

A plane-strain model taking into account the mechanics of long multilayer membranes with compliant supports has been developed before [32]. It was extended by including the mechanical characteristics arising from pressure loads [33]. Stacks of films with different prestress values necessarily lead to initial bending moments in the membrane, as evidenced by the integral and explicit forms of $\bar{S}_1$ in Table 3.1. As a result, the full form of Eq. (17) has to be considered for the clamping conditions. Under these circumstances, the normalized solution of Eqs. (14) is given by

$$\bar{w}(\bar{x}) = \frac{\bar{p}}{2\bar{S}}\left(\frac{1}{4} - \bar{x}^2\right) - \left[\frac{\bar{p}}{2\bar{S}}(2+\bar{K}) + \bar{S}_1\right]\frac{\cosh(\sqrt{\bar{S}}/2) - \cosh(\sqrt{\bar{S}}\bar{x})}{\bar{S}\cosh(\sqrt{\bar{S}}/2) + \bar{K}\sqrt{\bar{S}}\sinh(\sqrt{\bar{S}}/2)} \tag{55}$$

again conforming to the first normalization scheme of Section 3.2.3. In contrast to Eq. (53), the deflection profile given by Eq. (55) is no longer symmetric around $\bar{p} = 0$, as a consequence of the presence of initial bending moments inherent to multilayer diaphragms. Moreover, it can be shown with the present model that the effect of $\bar{S}_1$ is only observed with structures whose supports are compliant. Such a case is described in Figure 3.8: for sufficiently small values of $\bar{K}$, the buckling point can now occur at normalized $\bar{S}_0$ values far below $-4\pi^2$, depending on $\bar{S}_1$. Note that the sign of $\bar{S}_1$ depends on the stacking order of the set of layers. Another important feature is that the buckling transition of the deflection profile no longer occurs around $\bar{w}_0 = 0$, as emphasized by the dashed line in Figure 3.8, the locus of the bifurcation points being a function of S_1. The insets in Figure 3.8 show the profiles of the membrane at different pressures for two opposite prestress conditions.

Since the denominators, $\bar{S}\cosh(\sqrt{\bar{S}}/2) + \bar{K}\sqrt{\bar{S}}\sinh(\sqrt{\bar{S}}/2)$, are the same in Eqs. (53) and (55), and from the discussion in the previous section, one can conclude that the buckling line force $\bar{S}_{inst}$ given by Eq. (54) and its variation with the stiffness of the supporting edges still hold for elastically clamped multilayer membranes.

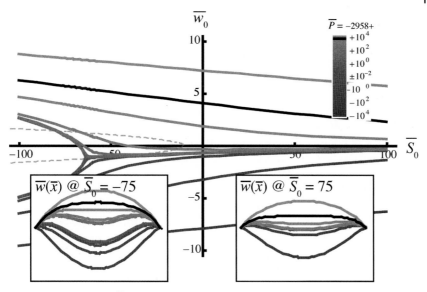

Figure 3.8 Isobaric $\bar{w}_0 - \bar{S}_0$ curves for simply supported devices ($\bar{K} = 10^{-2}$) with initial bending moments per unit length ($\bar{S}_1 = -300$). The insets show the deflection profiles of membranes with opposite values of $\bar{S}_0$ for several pressures $\bar{P}$.

As was discussed in the previous section, most of the devices commonly characterized using the bulge technique are mechanically equivalent to structures with rigid supports (or at most slightly compliant). In this way, the effect of $\bar{S}_1$ in the $\bar{w}_0$ versus $\bar{S}_0$ characteristic can be neglected and, as shown in Figure 3.7a, the assumption of the buckling point occuring at $\bar{S}_0 = -4\pi^2$ is reasonable, even for multilayers. In terms of device design rules, this brings some implications when dealing with multilayer diaphragms. From the application point of view, it is desirable to work with tensile (or weakly compressive), unbuckled membranes, which happens for $\bar{S}_0 > -4\pi^2$. From the normalization of $\bar{S}_0$ and the explicit form of S_0 in Table 3.1, this condition can be rewritten as

$$\sum_{n=1}^{N} (\sigma_{0,n} t_n) \Big/ \sum_{n=1}^{N} \left\{ E_{ps,n} t_n \prod_{l=1}^{2} \left[\sum_{m=1}^{n} t_m - (1 + i(-1)^l / \sqrt{3}) t_n / 2 - z_0 \right] \right\} > -4\pi^2 / a^2 \tag{56}$$

in terms of the plane-strain modulus, prestress and thickness of every single layer in the composite structure, which can then be tuned accordingly. The same reasoning can be applied as well for the reliable characterization of thin films with high levels of compressive prestress when using the bulge technique: as discussed in Section 3.4, such films can be stacked with highly tensile reference films with appropriate thickness so that the overall structure is tensile and the analysis is thus simplified.

3.3.7
Extraction of Fracture-mechanical Parameters

The bulge test can be used for the determination of the fracture properties of thin films, provided that both data with statistical significance and an accurate description of stress distributions within the diaphragms are available. On the one hand, a bulge setup has recently been fully automated for the measurement of all membranes in a wafer, realizing for the first time a high-throughput acquisition of mechanical thin-film data with convincing statistical control [33, 35]. The throughput is much higher than previously reported for this technique [5, 34, 57, 82], making the bulge test a suitable tool for the reliable characterization of both elastic and fracture properties of materials. On the other hand, the inclusion of compliant supports in the multilayer model described in Section 3.3.6, which already takes into account both stretching and bending stiffnesses of the suspended structures, enables a more accurate description of the stress distribution and, consequently, leads to a more precise extraction of fracture strengths for a general stack of films. In order to extract fracture properties from thin films, the membranes are pressure driven until fracture occurs. The stress distribution at the moment of failure within the diaphragms is computed as follows and used to analyze the brittle material strength via Weibull distributions [83, 84].

The stress distribution can be obtained from the deflection profile as defined by Eq. (55). The effective normalized (local) line force in layer i of the diaphragm is in this case given by

$$\bar{S}_{xx,i} = \bar{S}_{0,i} + \frac{1}{2}\frac{D_{0,i}}{D_0}\int_{-1/2}^{1/2}\left(\frac{d\bar{w}}{d\bar{x}}\right)^2 d\bar{x} - \frac{D_{0,i}}{D_0}(\bar{z}-\bar{z}_0)\frac{d^2\bar{w}}{d\bar{x}^2} \tag{57}$$

In the unloaded state, the membrane is prestressed with homogeneous line forces $\bar{S}_{0,i}$, as expressed by the first term in the right-hand side of Eq. (57). The second term corresponds to a stress increase arising from the elongation of the structure upon load application and the last term, proportional to the second derivative of the deflection, takes the thickness dependence into account by including bending effects. Note that the latter vanishes at the reference plane, that is, when $z = z_0$. In order to recover the stress distribution within a given layer of the membrane from Eq. (57), the same normalization as that of $\bar{S}$, $\bar{S} = Sa^2/D_2$, is used to obtain $S_{xx,i}$ and thus $\sigma_{xx,i}$:

$$\sigma_{xx,i} = \sigma_{0,i} + \frac{E_{ps,i}}{2a}\int_{-a/2}^{a/2}\left(\frac{dw}{dx}\right)^2 d\bar{x} - E_{ps,i}(z-z_0)\frac{d^2w}{dx^2} \tag{58}$$

In Eq. (58), the compliance of the supporting edges is implicitly considered through the selected reduced deflection function. An important fact to be considered when analyzing stress profiles at the moment of fracture is that the bending at the edges of the membrane leads to high local stresses with maximum tensile values occuring, according to Eq. (58), at the bottom of the diaphragm ($z = 0$).

The Weibull distribution function [83],

$$F = 1 - \exp\left[-\int_{V_+}\left(\frac{\sigma}{\sigma_{0,W}}\right)^m dV\right] \tag{59}$$

is widely used in the analysis of brittle material strength data [33, 82, 84]. In this equation, F is the probability of failure of a component subject to a spatially varying stress σ. The integral is taken over all elements dV within the volume V_+, where σ is tensile, i.e. $\sigma > 0$. In Eq. (59), the Weibull modulus m is a measure of the variability of the fracture stress (the higher the value of m the more predicable is the material) and $\sigma_{0,W}$ is the so-called scale parameter of the material, which has the awkward dimensions of $[\text{stress}][\text{volume}]^{1/m}$. For fixed values of m and V_+, the mean fracture stress increases with $\sigma_{0,W}$. The Weibull distribution builds on the weakest-link hypothesis [84] and therefore predicts a size effect: for a uniform volume distribution, the larger flaws are more likely to occur in larger specimens and larger specimens also have higher probabilities of having a certain size of flaws than smaller specimens. As a consequence, larger specimens are expected to be weaker. Another important fact is that fracture does not necessarily initiate at the highest stress locations, since a more severe flaw located at a sligthly less highly stressed region may be the first to become critical. As a result, when predicting failure loads it is necessary to take into account the probability of each region of the component; it is not sufficient to consider only the most highly stressed zone.

In the case of diaphragms under uniform applied differential pressure, the stress distribution σ given by Eq. (58) is not uniform, thus making the output of Eq. (59) difficult to interpret. The use of the standardized form of the Weibull distribution is more suitable in this situation [84, 85]. In the case of a tensile sample, the stress is uniform, $\sigma = \sigma_R$. By performing the integration over a reference volume of the tensile specimen of $V_+ = V_E$, Eq. (59) yields

$$F = 1 - \exp\left[-\left(\frac{\sigma_R}{\sigma_{0,W}}\right)^m V_E\right] \tag{60}$$

Comparing Eqs. (59) and (60) one sees that, when the condition $\sigma_R^m V_E = \int_{V_+}\sigma^m dV$ is satisfied, a given component has the same chance of failure as a tensile specimen subject to a stress σ_R. Therefore, if the stress at every point of the specimen is expressed as a multiple of some convenient reference stress σ_R, one can define the effective volume of the component as

$$V_E = \int_{V_+}\left(\frac{\sigma}{\sigma_R}\right)^m dV \tag{61}$$

It can be shown that the mean failure stress of components following this distribution is μ, where

$$\mu = \sigma_{0,W} V_E^{-1/m} \Gamma\left(\frac{m+1}{m}\right)$$

(62)

Combining Eq. (60) with Eq. (62) gives

$$F = 1 - \exp\left\{-\left[\frac{\sigma_R}{\mu} \Gamma\left(\frac{m+1}{m}\right)\right]^m\right\}$$

(63)

This standardized form of the Weibull distribution was first used by Robinson [84, 85]. It is particularly useful in discussing variability since it is true for any specimen geometry and loading arrangement.

For the case of multilayer diaphrams with compliant supports and under uniform pressure loads, the non-uniform stress distribution given by Eq. (58) makes Eqs. (61) and (63) most convenient for implementing Weibull statistics. If fracture originates in flaws uniformly distributed within the membrane volume, one can determine the reference stress σ_R for an arbitrary (tensile) reference volume V_E and from the failure stress distribution. The values of σ_R thus obtained from the measurement of a large enough set of samples can then be fit to Eq. (63) in order to extract both mean fracture strength μ and Weibull modulus m. The fitting procedure itself is an iterative/recursive process between Eqs. (61) and (63) as explained in detail in Section 3.4. If the material strength is dominated by line or surface rather than volume flaws, the volume integral in Eq. (59) can be replaced by line or surface integrals, respectively, with the subsequent analysis performed accordingly.

3.3.8
Thermal Expansion

Using Eqs. (43) and (50), the residual strain ε_0 of compressively prestressed square and long diaphragms with respect to the substrate can be inferred from their postbuckling amplitude. If such measurements are performed as a function of temperature, the derivative $d\varepsilon_0/dT$ is obtained. By definition, the coefficient of thermal expansion α_f of the thin film is

$$\alpha_f = \alpha_s + \frac{d\varepsilon_0}{dT}$$

(64)

where α_s denotes the corresponding coefficient of the substrate. This procedure was demonstrated with a PECVD silicon nitride film, where $\alpha_{SiN}(T) = \alpha_0 + \alpha_1(T - T_0)$ with $\alpha_0 = (1.803 \pm 0.006) \times 10^{-6} \, \text{K}^{-1}$, $\alpha_1 = (7.5 \pm 0.5) \times 10^{-9} \, \text{K}^{-2}$, and $T_0 = 25\,°C$ was found for the temperature range between 25 and 140 °C [41].

3.4
Experimental

3.4.1
Fabrication

The fabrication of membranes, when compared with that of test structures with different geometries, is a fairly simple process. It relies strongly on the anisotropic etching of silicon (Si) substrates by alkaline solutions such as ethylenediamine–pyrocatechol (EDP), potassium hydroxide (KOH) and tetramethylammonium hydroxide (TMAH) [86]. The anisotropy results from the different etch rates along the different crystallographic orientations. Etch rate ratios between the $\langle 100 \rangle$ and $\langle 111 \rangle$ directions as high as $35:1$, etch rates of the $\langle 100 \rangle$ plane up to $1.5\,\mu\text{m}\,\text{min}^{-1}$, along with high selectivities to silicon nitride (SiN_x) and silicon oxide (SiO_x) films [86, 87], make it possible to etch through full wafers in an inexpensive, batch-mode and reliable way in order to obtain suspended diaphragms. Deep reactive ion etching (DRIE) techniques, which have high etch rates for silicon with almost vertical side walls, have also been used to fabricate membranes [70].

A general fabrication process for suspended diaphragms is schematically shown in Figure 3.9. Prior to the silicon substrate etching, an etch mask layer is deposited on the rear of the silicon wafer whereas its front is coated with an etch stop layer, the stack of thin films to be mechanically characterized, and a protection coating, as shown in Figure 3.9a. After defining the rear etch mask, as shown in Figure 3.9b, the substrate is anisotropically etched (Figure 3.9c). The selective removal of both etch stop and protective layers (Figure 3.9d and e) is performed as the final step. The choice of etch mask, etch stop and protective layers and anisotropic chemical etchant depend on the respective etch selectivities to the mechanical layer

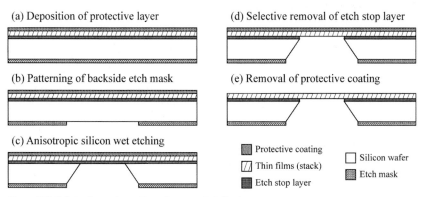

(a) Deposition of protective layer

(b) Patterning of backside etch mask

(c) Anisotropic silicon wet etching

(d) Selective removal of etch stop layer

(e) Removal of protective coating

 Protective coating
 Thin films (stack)
 Etch stop layer
 Silicon wafer
 Etch mask

Figure 3.9 Fabrication schematics of suspended diaphragms. (a) Initial configuration of substrate and layers. (b) The etch mask is patterned, followed by (c) the anisotropic silicon etching step. (d) The etch stop layer and (e) protective coating are selectively removed at the end of the process.

(or stacks of layers) to be characterized. For instance, in order to obtain polycrystalline silicon (polysilicon, poly-Si) membranes, an etch mask of stoichometric silicon nitride (Si_3N_4) grown by low-pressure chemical vapor deposition (LPCVD) and an etch stop layer made of LPCVD low-temperature oxide (LTO) can be used for stepwise KOH/TMAH steps [47]: the KOH solution is used to remove almost all the thickness of the silicon substrate, whereas the TMAH is chosen to finish the etching due to its higher selectivity to LTO. A previously spun poly(methyl methacrylate) (PMMA) film is used as the protective coating in order to prevent the TMAH solution from reaching adjacent structures in case some membrane breaks upon etching completion. The PMMA is then removed with acetone and the LTO etch stop layer can be dissolved with hydrofluoric acid (HF), without significant damage to the poly-Si film.

Other reported etch mask/etch stop layer/structural film/etchant configurations include SiN_x deposited by plasma-enhanced chemical vapor deposition (PECVD)/ PECVD SiN_x/PECVD SiN_x/KOH [6, 31], LPCVD Si_3N_4/stack of LPCVD Si_3N_4, titanium (Ti) and aluminium (Al)/low-k polymer/KOH [71], SiN_x/SiN_x/copper (Cu)/KOH [66], thermal oxide (ThOx)/ThOx/polyimide/hydrazine solution [11], silicon oxide/p^+ diffusion of boron/polymer/hydrazine [10] and LPCVD Si_3N_4/ LPCVD Si_3N_4/multilayers of nitrides and oxides/KOH–TMAH [33]. In the last case, LPCVD Si_3N_4 layers are used, for two reasons: (i) these films have extremely high values of tensile stress (above 1 GPa) [33–35] and therefore are ideal for stress compensation of stacks including compressive films, and (ii) the etch rate of Si_3N_4 films by KOH is extremely low (below 0.9 nm h^{-1} for 30 wt.-% KOH at 85 °C) when compared with other nitrides or oxides [34, 35, 86, 87]. A thin layer of Si_3N_4 can endure sufficiently long etching during the bulk micromachining of Si frames for the definition of the diaphragms, making it an excellent rear etch mask. The most commonly reported geometries include square and rectangular diaphrams with side lengths ranging from below 100 μm to a few mm [1, 5, 10, 22, 31, 33, 34, 47, 66, 70, 82].

3.4.2
Measurement Techniques

In the first bulge experiments, the deflection of circular membranes made of gold (Au) and silver (Ag) was measured by optical interference means [1]. Since then, advances in automation and characterization tools have led to significant improvements of the bulge technique.

While the application of a controlled differential pressure to a membrane is straightforward and easily implemented, care has to be taken concerning the measurement of the deflection. Contact measurements have been performed using mechanical profilometers [44]. Even though current profilometers possess good resolution over large measurement ranges, the approach is not well suited for this purpose since the contact stylus inevitably influences the overall membrane response. In contrast, due to their non-contact and non-destructive nature, optical techniques are preferred and have been widely used to measure the profiles of bulged diaphragms. One optical approach is based on interference: a flat cover glass is placed over the test area, in close proximity to the diaphragm. Upon monochromatic illumination, the deflection can be evaluated from the resulting interference pattern (Newton rings) [1, 15, 25, 28, 51, 66, 71]. Such a procedure can be automated via digital image analysis [15, 25, 28]. Another way is to focus a laser obliquely onto the membrane and measure the deflection of the reflected beam using position sensitive photodetectors [65]. Several authors have simply used microscopes with calibrated vertical movement, where the profile amplitude is measured by focusing the optical instrument on top of the structure [10, 11, 56]. Particular optical sensors, able to auto-focus themselves on the membrane and thus to measure the profile automatically, have been used by the authors [6, 31, 33–35, 39, 47].

A truly wafer-scale bulge setup has recently been developed by the authors [33, 35, 47]. Figures 3.10 and 3.11 show a photograph and the schematics, respectively, of the experimental apparatus prepared for the full wafer measurement of membranes under uniform differential pressure. The sensing part of the setup consists of an AF16 Hyperion system (OPM Messtechnik, Ettlingen, Germany), which includes an optical auto-focus sensor capable of measuring profiles up to $1500\,\mu m$ with a resolution down to 10 nm and a motorized x–y table composed of two perpendicular linear stages covering an overall travel range of $150 \times 100\,mm$ and in-plane resolution of $0.2\,\mu m$. Point, line and area scans are possible with this system. Additionally, a rotation stage has been included for alignment purposes. The actuation component consists of a DPI 520 pressure controller (Druck, Leicester, UK) that pressurizes several terminals of solenoid valves (Festo Esslingen, Germany) with values ranging from –80 to 800 kPa. The pneumatic valves are controlled using relays cards (QUAN34-COM Informationssysteme, Wesseling, Germany), that redirect the pressure to the membranes via appropriate lines,

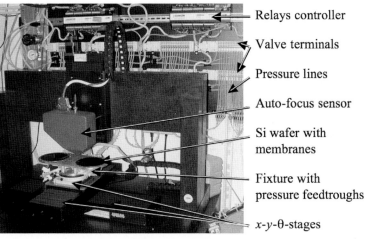

Relays controller

Valve terminals

Pressure lines

Auto-focus sensor

Si wafer with membranes

Fixture with pressure feedtroughs

x-y-θ-stages

Figure 3.10 Automated bulge test setup for full wafer measurements.

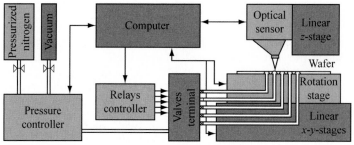

Figure 3.11 Schematics of automated bulge test setup. A controller adjusts the pressure value, which is redirected to a given membrane via relay-controlled solenoid valves. An optical sensor measures the resulting deflection profile. The use of automated linear stages enables the acquisition of point, line and area scans.

feedthroughs and a wafer fixture. The setup is programmed for the automated sequential characterization of 80 diaphragms of a wafer mounted on one chuck. A more flexible configuration with 16 valves is available at a second chuck of the fixture.

3.4.3
Procedures

The procedures to extract both elastic and fracture properties of thin films from the data obtained using the setup above are now described. For illustration, the method is applied in this section to materials with tensile and compressive pre-stress, i.e. LPCVD Si_3N_4 and ThOx, respectively. Rectangular geometries, with aspect ratio 1:10 and widths ranging from 400 to 800 μm, are adopted so that the simpler plane-strain model described in Section 3.3.5 can be used. As the nitride films present tensile residual stresses, single-material diaphragms suffice whereas the characterization of the compressively stressed oxide films has to be performed by stacking these films with stress-compensating layers, LPCVD Si_3N_4 in this example, so that Eq. (56) is verified.

The deflection profile of the extended middle section of each membrane is measured for increasing values of applied pressure until fracture occurs. This routine is sequentially performed until all 80 membranes on the wafer have been measured. Batch data post-processing routines are then applied to
- level every raw profile
- extract the membranes dimensions
- obtain the center deflection as a function of pressure
- perform the data fits for every membrane.

Figure 3.12a shows the levelled profiles obtained for different pressures of a 6045 μm long, 604 μm wide and 102 nm thick LPCVD Si_3N_4 membrane. The center

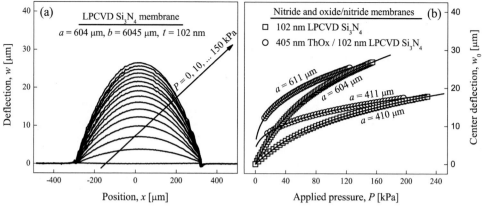

Figure 3.12 (a) Levelled deflection profiles of LPCVD Si₃N₄ membranes obtained at different pressure loads. (b) Center deflection vs. pressure of single and bilayer membranes with different sizes. For each data set, the highest pressure point corresponds to the moment prior to the failure of the diaphragm. The lines are fits to the plane-strain model given by Eq. (55).

deflection of single LPCVD Si₃N₄ structures and composite diaphragms of Si₃N₄ on top of 405 nm of ThOx, as a function of pressure, are shown in Figure 3.12b, for targeted widths of 400 and 600 μm. The lines are fits to Eq. (55), from which both plane-strain modulus E_{ps} and prestress value σ_0, are extracted. The agreement between the fits and experimental data from membranes with different sizes emphasizes the correct dimensional scaling predicted by the model. The histograms of E_{ps} and σ_0 values obtained for LPCVD Si₃N₄ and thermal oxide (using an Si₃N₄ reference layer) from full wafer measurements are presented in Figure 3.13. The values of E_{ps} = 279 ± 11 GPa (59 ± 5 GPa) and σ_0 = 1191 ± 31 MPa (−318 ± 15 MPa) extracted for the nitride (oxide) are in agreement with literature values for similar films [15, 26, 31, 34, 48] ([48, 52]) but, in contrast, the uncertainty windows result from significantly larger numbers of samples.

The membranes are bulged until fracture occurs. The failure pressure is recorded and, along with the E_{ps} and σ_0 values extracted for the various films in the stack and for each membrane, the stress distribution is evaluated by means of Eq. (58). In order to illustrate the stress distribution at the fracture load within the diaphragms as a function of the coordinate perpendicular to the diaphragm plane, calculated at the center ($x = 0$) and edges ($x = \pm a/2$) of the diaphragms, are plotted in Figure 3.14a, top and bottom plots, respectively. The 411 μm wide, single layer LPCVD Si₃N₄ membrane (with data represented by the dash-dotted lines) experiences a tensile stress value of about 2.7 GPa at the diaphragm center, with almost no z dependence. At its edges, however, the stress peaks at strongly tensile and compressive levels, at the bottom and top of the structure, respectively. The

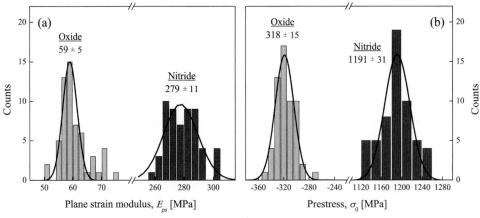

Figure 3.13 Histograms of (a) plane-strain modulus and (b) prestress obtained from full wafer bulge measurements for thermally grown silicon oxide and LPCVD silicon nitride.

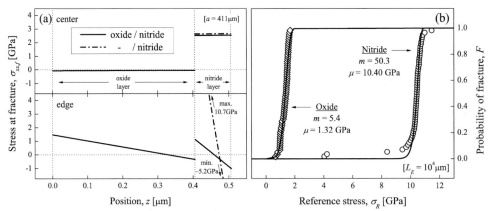

Figure 3.14 (a) Dependence of the stress in a diaphragm as a function of the coordinate perpendicular to the diaphragm at the moment of fracture, evaluated at $x = 0$ and $x = \pm a/2$, top and bottom plots, respectively. The dash-dotted and solid lines correspond to single and bilayer 411 μm wide diaphragms, respectively. (b) Standardized Weibull data and fits for the nitride and oxide layers.

membrane experiences a maximum tensile stress of 10.7 GPa at $x = \pm a/2$ and $z = 0.41$ μm, i.e. along its lower edge. This provides a measure of the nitride strength. The failure stress distributions obtained for an oxide/nitride diaphragm with comparable width are plotted in Figure 3.14a as solid lines. One immediate observation is that the maximum tensile stress occuring at the top-nitride layer in the oxide/nitride stack, 2.6 GPa, is far lower than the values that these films can withstand in the single-layer configuration. It can then be concluded that the fracture in the bilayers does not originate at the LPCVD Si_3N_4, the oxide layer consequently being the weakest link responsible for the failure of the device. As an indication of the ThOx strength, the maximum stress experienced by the oxide in bilayers is 1.46 GPa at $x = \pm a/2$, $z = 0$ (oxide bottom edges).

From the stress distributions within the diaphragms at the moment of fracture, the brittle material strength is extracted by fitting the experimental data with Weibull distributions. As mentioned in Section 3.3.7, such fits are based on an iterative scheme involving Eqs. (61) and (63) [34]. The reference tensile stress is given by Eq. (61).

The stress distribution $\sigma_{xx,n}$ given by Eq. (58) is not valid for the full portion of the diaphragms, in particular near its short edges. However, and in order to take advantage of the plane-strain situation occuring in the extended middle section of the diaphrams, that expression may be inserted into the integral of Eq. (61), which is then reduced to a two-dimensional form. Its evaluation along the long edges in the range $-b/2 \leq y < b/2$, which yields L, is corrected to $L - ca$ [34]:

$$\sigma_R = \left(\frac{L - ca}{V_E}\right)^{1/m}\left[\int_{S_+} \sigma^m_{xx,n}(x, z)dxdz\right]^{1/m} \tag{65}$$

This integral is now computed over the cross-sectional area S_+ of the extended middle section of the membrane where $\sigma_{xx,n} > 0$. The inclusion of the factor c is intended to correct for corner effects and for the stress distributions along the shorter edges of the structure. Nevertheless, it has been demonstrated previously that the value of c does not significantly affect the strength μ and Weibull modulus m fit results and also the quality of the fit and therefore a choice of $c = 0$ is considered reasonable [34]. In case fracture originates at flaws uniformly distributed on a surface or along a line, similar reasoning leads to

$$\sigma_R = \left(\frac{L - ca}{A_E}\right)^{1/m}\left[\int_{L_+} \sigma^m_{xx,n}(x, z = z_S)dx\right]^{1/m} \tag{66}$$

and

$$\sigma_R = \left(\frac{L - ca}{L_E}\right)^{1/m} \sigma_{xx,n}(max) \tag{67}$$

where A_E and L_E are equivalent tensile areas and lengths, and z_S denotes the location of the surface or interface suspected to be responsible for the failure. Note that the integration in Eq. (66) is carried out over the positive-stress region L_+ of the membrane edges across the thickness of the weaker layer and $\sigma_{xx,n}(max)$ in Eq. (67) is the maximum fracture tensile stress occuring at the layer being analyzed. In LPCVD Si_3N_4 thin films, volume defects resulting from the amorphous structure may contribute to the failure, surface flaws can arise from processing conditions or line singularities along the film/substrate interface may dominate the fracture mechanism [34]. From the size effect predicted by the Weibull analysis

[34, 83, 84], the ratios between the equivalent tensile strength of specimens with different volumes, areas and lengths are $\mu_1/\mu_2 = (V_{E,1}/V_{E,2})^{-1/m}$, $\mu_1/\mu_2 = (A_{E,1}/A_{E,2})^{-1/m}$ and $\mu_1/\mu_2 = (L_{E,1}/L_{E,2})^{-1/m}$, respectively. By measuring samples with different sizes/ dimensions, the occurrence of one these conditions can be verified and a conclusion about the fracture-causing mechanism can in principle can be drawn. Such a procedure has been previously applied to the fracture study of LPCVD Si_3N_4 membranes, from which marginally better fits were obtained for line flaws [34].

Figure 3.14b shows the pooled Weibull data (i.e. data obtained from samples with different dimensions and normalized to the same reference size) and respective fits to Eq. (63) for nitride membranes and oxide layers in oxide/nitride diaphragms, normalized to a length L_E of 104 μm. The LPCVD Si_3N_4 films present a relatively high fracture strength value of 10.4 GPa, in agreement with previous results [34] and Weibull modulus m of 50.3. The latter is higher than previously reported from bulge tests, which is thought to be due mainly to the higher throughput of experimental data and the superior experimental control achieved with the wafer scale bulge setup. The oxide data reveal it as a weaker and much less predictable material as designated by μ = 1.32 GPa and the low Weibull modulus m = 5.4. To the authors' knowledge the fracture characterization of oxides using the bulge technique has never been reported before.

3.5
Results

A compilation of elastic and fracture mechanical properties of several MEMS thin-film materials, extracted from bulge experiments, is presented here. This section is divided into four main parts, concerned with (i) silicon nitride, (ii) silicon oxide, (iii) polysilicon and (iv) metals, polymers and other materials. One goal is to give a broad impression of the range of values that the properties can cover, depending on processing conditions. The applications of the various thin films are briefly discussed along with the influence of their mechanical properties on the layer functionality. A second aim is to provide the reader with a reliable, compact and as complete as possible database of mechanical properties to be considered by microsystems engineers during device design, fabrication and characterization. Apart from the fracture-mechanical properties of silicon, little is known about the strength and predictability of other brittle materials for MEMS: these data are now presented in the following tables, with a large fraction of them having been obtained using the wafer-scale bulge test setup described in the previous section.

3.5.1
Silicon Nitride

Silicon nitride thin films have a relevant role in semiconductor industry: these layers form an effective diffusion barrier in the local oxidation of silicon and provide excellent mechanical and chemical passivation of integrated circuits. In addition, silicon nitride films are used as masks in semiconductor processing, wear-resistance coatings for tools and electrode insulation [57]. A low-temperature deposition technique such as PECVD is necessary to prevent undesired interdiffusion of adjacent layers [57] and unintended changes of diffusion profiles. Diaphragms and membranes of silicon nitride are commonly used in MEMS accelerometers and pressure sensors [6, 31, 32, 48, 51, 57, 58, 75, 82]. With excellent chemical stability in alkaline solutions [57, 82, 86, 87], high strength and high values of the elastic modulus [34, 35] and insensitivity of elastic and fracture characteristics up to 1050 °C [88], LPCVD Si_3N_4 is used in MEMS applications where high temperature and harsh environments preclude the use of silicon-based sensors [82].

The data displayed in Table 3.3 show that the mechanical properties of nitride thin films depend strongly on the deposition method and, for a given deposition technique, these are largely influenced by the deposition parameters. For the PECVD dielectrics, the reactor frequency and inclusion of hydrogen greatly affect the prestress level. The residual stress can vary from −1274 MPa (extremely compressive) to 412 MPa (tensile) when the frequency changes from 187 kHz (low frequency, LF) to 13.56 MHz (high frequency, HF). Systems such as the STS Mesc Multiplex PECVD reactor (Surface Technology Systems, Newport, UK) take advantage of this fact by using recipes with alternating frequency (mixed frequency, MF) steps in order to tailor the film stress for a particular application. This fact is illustrated by the wide range of prestress values of MF PECVD nitride thin films listed in Table 3.3. The PECVD silicon nitrides are usually grown from silane (SiH_4) and ammonia (NH_3) mixtures at temperatures of at least 300 °C in order to obtain good-quality films, i.e. with low pinhole densities [89]. The plane-strain modulus of these films is not so strongly affected by the reactor frequency. The LPCVD nitrides are also obtained from the dissociation of SiH_4 (or alternative Si containing percursor) and NH_3 gases, but at temperatures between 700 and 800 °C [89, 90]. For these films, the deposition gas flow ratio between the Si and N precursors greatly influences the prestress values and a distinction between silicon-rich and nitride-rich films is often made [90]. The bulge data in Table 3.3 show that silicon-rich LPCVD films have tensile residual stresses much lower (114–226 MPa) than the nitride-rich layers, which can reach values up to 1194 MPa. Plane-strain moduli of the LPCVD films of around 280 GPa are almost a factor of 2 larger than those of the PECVD layers, with no visible dependence on silicon/nitride contents.

The Weibull fracture parameters obtained from the bulge data of several nitrides are given in Table 3.4. As a reference for the statistical quality of the data, the number of membranes measured for each case is also presented. Note that the largest possible number of specimens should be used to estimate Weibull parameters since a lower count of specimens translates into increased uncertainty in the true value of both strength and Weibull modulus. From the data of the various sources, it is manifest that the LPCVD films present much higher strengths and Weibull moduli than the PECVD layers. The difference in the strengths may

Table 3.3 Elastic modulus E or plane-strain modulus $E_{ps} = E/(1 - v^2)$ and prestress σ_0 obtained from the bulge testing of silicon nitride films grown by various methods and under different conditions.

E, E_{ps} (GPa)	σ_0 (MPa)	Deposition conditions and remarks[a]
E_{ps} = 166 ± 29, 143 ± 12, 170 ± 23	−1274 ± 64, −323 ± 25 and 412 ± 28	PECVD (LF, MF and HF, 330 °C), 0.1–1 μm, 1:10 and 1:20 membranes, plane-strain single layer and multilayers [33]
E = 150 ± 7	−320 ± 15	PECVD (MF, 300 °C), 1.9 μm thick square membranes, 101–3050 μm wide, buckling symmetry transitions [39]
E = 160	$−1.78 \times 10^{-3} E/(1 - v)$	PECVD (MF, 300 °C), 0.49 μm, rectangular 1:6 membranes, 0.1–1 mm wide, symmetry transitions [37]
E_{ps} = 134.4 ± 3.9, 142 ± 2.6	1.3 ± 3.8, −63.2 ± 12.4	PECVD (MF), 3.5 and 0.7 μm, weakly tensile and compressive long membranes, adjusted by LF tuning [31]
E_{ps} = 142.6 ± 1.4 (v = 0.253 ± 0.017)	−67.2 ± 8.3	PECVD (MF), ripple periods variation [38]; v = 0.235 ± 0.04 from combined long and square membranes [6]
E = 132	−44 ± 5, −24 ± 5	PECVD, 0.6 μm, as deposited and after 400 °C annealing step, Infineon [49]
E = 97 ± 6	82 ± 6	PECVD, sensor passivation, ALP2LV, EM[b] [48]
E = 210	110	PECVD (300 °C), 0.1–0.3 μm thick, 2 × 2 and 2 × 8 mm² suspended diaphragms [15]
E_{ps} = 145.5 ± 3.8	111.1 ± 18	PECVD, 0.36 μm thick long membranes, shorter edge lengths from 0.6 to 1 mm, plane-strain regime [59]
E = 276	226 ± 2	LPCVD (835 °C), silicon-rich films [56]
E = 255 ± 5	114–130	LPCVD, 0.5 μm, low-stress films, square 1-mm² membranes, measured at Johns Hopkins University and Exponent [60]
E = 260, 320	NA	LPCVD (850 °C), as deposited and after 1050 °C anneal [26]
E = 290	1000	LPCVD (790 °C), 0.1–0.3 μm thickness range, 2 × 2 and 2 × 8 mm² suspended diaphragms [15]
E_{ps} = 288	1040 ± 160	LPCVD, 73 and 83 nm thick, 1:6 and 1:25 membranes, shorter lengths from 0.2 to 0.6 mm, plane strain [34]
E_{ps} = 278 ± 10	1194 ± 30	LPCVD (770 °C), 50–250 nm, 1:10 and 1:20 membranes, 0.3–0.8 mm wide, automated wafer level bulge setup [33, 35]
E_{ps} = 279 ± 9	1070 ± 25	LPCVD (780 °C), 145 nm, rectangular diaphragms, 0.65–1.05 mm wide, deposited at Infineon, measured at IMTEK [61]

a   Abbreviations: low (LF, 187 kHz), mixed (MF) and high frequency (HF, 13.56 MHz), plasma-enhanced chemical vapor deposition (PECVD) and low-pressure chemical vapor deposition (LPCVD). NA, not applicable.

b   2-μm low-voltage CMOS process ALP2LV from EM Microelectronic-Marin (EM).

Table 3.4 Fracture strength μ and Weibull modulus m obtained from the bulge testing of different silicon nitride thin films. The number of membranes measured is also presented.

μ (GPa)	m (No. of membranes)	Deposition conditions and remarks[a]
1.32[b]	8.2 (56)	PECVD (MF, 330°C), 0.1–1 μm, 1:10 and 1:20 membranes, plane-strain regime achieved with bilayers [33]
1.18[b]	7.3 (55)	PECVD (HF, 330°C), 0.1–1 μm, 1:10 and 1:20 membranes, 0.3–0.8 mm wide, plane-strain regime [33]
0.39 and 0.42	NA (17) and NA (14)	PECVD (HF, 250°C), silicon- and nitrogen-rich 1-μm films, circular freestanding structures, 1-cm radius [57]
0.9[b]	NA (NA)	LPCVD (850°C), 1-μm silicon-rich films, circular membrane specimens, 1-cm diameter [5]
5.69 ± 0.25	NA (7)	LPCVD, 0.5 μm, low-stress films, square 1 mm² membranes, measured at Johns Hopkins University and Exponent [60]
2.6	4.7 (20)	LPCVD, 0.2 μm, highly tensile films, square membranes with widths ranging from 1.03 to 1.11 mm [58]
8.5[c]	61.1 (68)	LPCVD (780°C), 145 nm, rectangular diaphragms, 0.65–1.05 mm wide, deposited at Infineon, measured at IMTEK [61]
10.4[b]	50.2 (579)	LPCVD (770°C), 50–250 nm, long 1:10, 1:20 membranes, 0.3–0.8 mm wide, automated wafer level bulge setup [33, 35]
12.1[b]	18.3 (52)	LPCVD, 73 and 83 nm thick, 1:6 and 1:25 membranes, shorter lengths from 0.2 to 0.6 mm, plane strain [34]

a Abbreviations: low (LF, 187 kHz), mixed (MF) and high frequency (HF, 13.56 MHz), plasma-enhanced chemical vapor deposition (PECVD) and low-pressure chemical vapor deposition (LPCVD). NA, not applicable.

b Data normalized to an equivalent tensile reference length $L_E = 10^4$ μm.

c Maximum stress calculation based on bent wafer technique (stress as a function of substrate bending height).

tentatively be explained by the lower hydrogen content and higher density of the LPCVD thin films [6, 89, 90]. This is even verified with the data obtained for the different films using the same full wafer setup, removing any possible doubt about the statistical quality of the data. Maximum fracture strengths of 1.32 GPa have been reported for the PECVD films with Weibull moduli below 10. In contrast, the strength of LPCVD Si_3N_4 is almost one order of magnitude higher, up to 12.1 GPa, and the bulge data show that this is a highly predictable material, as evidenced by the extremely high values of m, up to 61.1. These values along with the relatively high elastic moduli and tensile prestresses (Table 3.3) make this material very interesting for MEMS applications. Examples where such mechanical properties of the LPCVD films are relevant include stress compensation in composite membranes [15, 33], high-performance micromechanical resonators [91] and high-temperature devices [82, 89, 92].

3.5.2
Silicon Oxide

Similarly to silicon nitride, oxides find applications in masking, isolation, passivation and planarization: membranes in lithographic fabrication, dielectric gates or insulators for metal–insulator–semiconductor devices, barrier coatings on polymer components and interfacial films in ceramic composites are some of the examples [49, 89, 93]. Due to the high selectivity of hydrofluoric acid solutions when etching silicon oxide with respect to silicon, along with relatively high etch rates, silicon oxide is commonly used as a sacrificial layer in micromachining processes [29, 89]. In addition, dielectric micromachined membranes on silicon are widely used in microcomponents, microsensors and microactuators as deformable mechanical parts, for thermal and electrical insulation and for the reduction of RF dielectric losses [29, 89]. The mechanical characterization of these films is also of great importance. The residual stresses, whatever their origin, can create serious problems in semiconductor technology where layers of differing materials must be grown on a silicon substrate: stress may cause crystal defects in the underlying substrate, cracking of deposited films and void formation in aluminum conductor lines, leading to reduced device yield [49, 94].

Films of doped or undoped silicon oxide can be grown or deposited using a variety of chemical reactions and physical conditions. For example, SiO_2 can be grown on a silicon surface at temperatures between 900 and 1200 °C in pyrolytic H_2O or dry oxygen, deposited on a substrate at temperatures as low as 200 °C from a plasma of silane and nitrous oxide (N_2O) by PECVD or deposited at 400–800 °C from the reaction of silane and oxygen by LPCVD. The decomposition of selected organosilicon esters and anhydrides at reduced pressures can also produce oxide films and it is often possible to dope these films with several percent of boron and phosphorus [49, 89, 95].

Both elastic and fracture properties of oxides prepared in different ways show some variation, as illustrated by the data, extracted from bulge measurements, in Tables 3.5 and 3.6. However, and in contrast to silicon nitride, the mechanical properties of thin oxide films do not vary as much with processing conditions, even for very distinct deposition processes. The data show a consistent values averaging around 60 GPa for the plane-strain modulus of silicon oxide films, most of them having compressive residual stress with values of σ_0 down to −395 MPa. The strength of the oxides, in Table 3.6, is in general lower than that of nitride films, in Table 3.4. The oxide films present rather low Weibull moduli, between 5.1 and 10.1, and therefore are less predictable from the fracture point of view. Even though silicon oxide is typically included in MEMS and microelectronics

Table 3.5 Elastic modulus E or plane-strain modulus $E_{ps} = E/(1 - v^2)$ and prestress σ_0 obtained from the bulge testing of silicon oxide films produced by various methods and under different conditions.

E, E_{ps} (GPa)	σ_0 (MPa)	Deposition conditions and remarks[a]
$E_{ps} = 76 \pm 8$	-395 ± 13	SiO$_x$, PECVD (330°C, LF), 0.47 μm, 1:10, 1:20 membranes, 0.3–0.8 mm wide, plane-strain stack, automated setup [33]
$E = 89$	200	Si$_x$O$_y$N$_z$, PECVD (LF), square diaphragms, mechanical properties studied as a function of the O/Si atomic ratio [29]
$E_{ps} = 72 \pm 11$	-312 ± 16	SiO$_x$, PECVD (330°C, HF), 0.47 μm, 1:10, 1:20 membranes, 0.3–0.8 mm wide, plane-strain stack, automated setup [33]
$E_{ps} = 104 \pm 10$	-94 ± 10	SiO$_x$, PECVD (380°C), 0.8 μm, PECVD SiO$_x$/LPCVD Si$_3$N$_4$ composite membranes with square geometry [51]
$E = 60$	-96 ± 9, -68 ± 5	Silane PECVD oxide, 0.7 μm, as deposited and after 30 min annealing at 400°C, Infineon [49]
$E = 60$	-103 ± 24, -96 ± 21	TEOS PECVD oxide, 0.7 μm, as deposited and after 30 min annealing at 400°C, Infineon [49]
$E = 20 \pm 10$	-40 ± 10	BPSG, ALP2LV, EM[b] [48]
$E = 57$	-27 ± 3, -26 ± 3	BPSG ox., CVD, 1 μm, 4% B and 4% P doping, as deposited and after 30 min annealing at 400°C, Infineon [49]
$E = 65.5 \pm 5$ ($v = 0.2 \pm 0.2$)	-37 ± 6	PSG, ALP2LV, EM[b] [48]
$E_{ps} = 53 \pm 8$	-321 ± 15	LTO, LPCVD (425°C), 0.43 μm, 1:10, 1:20 membranes, 0.3–0.8 mm wide, plane-strain stack, automated setup [33]
NA	-300	FOX, ALP2LV, EM[b] [48]
$E = 70$	-237	ThOx, Si oxidation (950°C), 1 μm, 1:6 diaphragms, up to 0.8 mm wide, stacked with tensile LPCVD Si$_3$N$_4$ [52]
NA	-300	ThOx, Si oxidation, 1 μm thick layers, stacked with c-Si (12–15 μm), long diaphragms, 100–5000 μm width [44]
$E_{ps} = 60 \pm 5$	-318 ± 15	ThOx, Si oxidation (950°C), 0.4 μm, 1:10, 1:20 membranes, 0.3–0.8 mm wide, plane-strain stack, automated setup [33]

a Abbreviations: low (LF, 187 kHz) and high frequency (HF, 13.56 MHz), plasma-enhanced (PE) chemical vapor deposition (CVD) (PECVD), low-pressure chemical vapor deposition (LPCVD), tetraethyl orthosilicate (TEOS), borophosphosilicate glass (BPSG), phosphosilicate glass (PSG), low-temperature oxide (LTO), field oxide (FOX) and thermal oxide (ThOx). NA, not applicable.

b 2-μm low-voltage CMOS process ALP2LV from EM Microelectronic-Marin (EM).

Table 3.6 Fracture strength μ and Weibull modulus m obtained from the bulge testing of different silicon oxide thin films. The number of membranes measured is also presented.

μ (GPa)	m (No. of membranes)	Deposition conditions and remarks[a]
0.49, 0.49, 0.53[b]	10.1 (63)	SiO$_x$, PECVD (330 °C, LF), 0.47 μm, 1:10, 1:20 membranes, 0.3–0.8 mm wide, plane-strain stack, automated setup [33]
0.55, 0.33, 0.36[b]	5.8 (40)	SiO$_x$, PECVD (330 °C, HF), 0.47 μm, 1:10, 1:20 membranes, 0.3–0.8 mm wide, plane-strain stack, automated setup [33]
0.26, 0.23, 0.21[b]	5.1 (71)	LTO, LPCVD (425 °C), 0.43 μm, 1:10, 1:20 membranes, 0.3 to 0.8 mm wide, plane-strain stack, automated setup [33]
0.8–1.48[c]	8 (40)	ThOx, Si oxidation (950 °C), 1 μm, 1:6 diaphragms, up to 0.8 mm wide, stacked with tensile LPCVD Si$_3$N$_4$ [52]
1.32, 0.37, 0.35[b]	5.4 (58)	ThOx, Si oxidation (950 °C), 0.4 μm, 1:10, 1:20 membranes, 0.3–0.8 mm wide, plane-strain stack, automated setup [33]

a   Abbreviations: low (LF, 187 kHz) and high frequency (HF, 13.56 MHz), plasma-enhanced (PE) chemical vapor deposition (CVD) (PECVD), low-pressure chemical vapor deposition (LPCVD), low-temperature oxide (LTO) and thermal oxide (ThOx).
b   Data normalized to an equivalent tensile reference length $L_E = 10^4\,\mu$m, area $A_E = 5 \times 10^6\,\mu$m^2 and volume $V_E = 2.5 \times 10^5\,\mu$m^3.
c   Data normalized to an equivalent tensile reference length $L_E = 10^4\,\mu$m.

processes, its use as a structural component should always be considered with care because of both its low stiffness and strength.

3.5.3
Polysilicon

Polysilicon thin films are commonly employed in integrated circuits to form gate electrodes, interconnects, emitters and other elements [46, 49, 96]. These films are also used as structural materials in a wide variety of MEMS applications. The possibility of obtaining reliable, low-stress films [89] allows its use in applications such as microgears, micromechanical resonators, relays, membranes and pressure sensors [46, 89, 96].

Silicon is deposited by thermal decomposition of silane, $SiH_4 \rightarrow Si + 2H_2$. Amorphous films are obtained when this decomposition is performed at temperatures below 580 °C. At even lower temperatures (300 °C and below), amorphous silicon can be grown by PECVD: these films are usually under compressive stress, down to −500 MPa, mostly because of their high content of hydrogen [89]. Between 580 and 610 °C depositions, fine-grained polysilicon nucleation takes place and highly tensile films with stresses above 1 GPa occur within this narrow range of temperatures [89]. Above 610 °C, larger grains form and grow with a columnar structure. Polysilicon thin films are normaly deposited by LPCVD of silane between 580 and 650 °C [89]. The use of the LPCVD technique is justified by its compability with high-volume wafer fabrication [46, 97]. As deposited, typical polysilicon films are usually under considerable stress, which can be annealout out at 1000–1100 °C [89, 98]. Doping of these films is possible, which has a strong effect on grain size and internal stress [49, 89].

Mechanical properties of LPCVD polysilicon films, obtained from bulge measurements, are listed in Table 3.7. In general, the elastic moduli of the different sources agree with each other. The spread in the prestress values, from –440 to 161 MPa, is attributed to differences in the processing conditions, namely gas flow

Table 3.7 Elastic modulus E or plane-strain modulus $E_{ps} = E/(1 - v^2)$ and prestress σ_0 obtained from the bulge testing of polysilicon films grown by low-pressure chemical vapor deposition (LPCVD) under different conditions.

E, E_{ps} (GPa)	σ_0 (MPa)	Deposition conditions and remarks[a]
$E = 192$	–440	Standard, 1×1, 2×2, 1×4 and $2 \times 8\,\mathrm{mm}^2$ diaphragms, tensile stacks with LPCVD Si_3N_4 [24]
$E = 151, 162$	–350–20	$T_{dep} = 620\,°C$, $0.46\,\mu m$, undoped, as deposited and after 2 h annealing at 1100 °C, stacked with LPCVD Si_3N_4 [25]
$E = 160$	–180	$T_{dep} = 630\,°C$, undoped, $0.2\,\mu m$, 2×2 and $2 \times 8\,\mathrm{mm}^2$ suspended composite diaphragms with LPCVD Si_3N_4 [14, 15]
$E_{ps} = 160 \pm 16$	-8 ± 16	$T_{dep} = 625\,°C$, 0.53 and $1.91\,\mu m$, annealed at 1050 °C, $1:10$, 0.4–0.8 mm wide diaphragms, automated setup [47]
$E = 188 \pm 14.6$ to 239 ± 13.5	25 ± 5.1 to 45 ± 9.9	$T_{dep} = 630\,°C$, $4\,\mu m$, annealed at 1000 °C, $3 \times 3\,\mathrm{mm}^2$ membranes, integrity studies from exposure to HF solutions [50]
$E = 162 \pm 4$ ($v = 0.20 \pm 0.03$)	103 to 161	$T_{dep} = 580\,°C$, $3.56\,\mu m$, annealed at 1050 °C, 2×2 and $6 \times 1.2\,\mathrm{mm}^2$ structures, MUMPS-12 run [46]
$E = 205, 210$	–160–32	$T_{dep} = 630\,°C$, $0.2\,\mu m$, B and P ion implanting (doses up to $10^{15}\,\mathrm{cm}^{-2}$) plus 900 °C annealing, $2 \times 8\,\mathrm{mm}^2$ tensile stacks [17]
$E = 52 \pm 6$ ($v = 0.2 \pm 0.1$)	-134 ± 10	n^+-Poly1, ALP2LV, EM[b] [48]
$E = 169$	$19 \pm 718 \pm 7$	$0.5\,\mu m$, n-doped $10^{16}\,\mathrm{cm}^{-3}$, as deposited and after 30 min annealing at 400 °C, Infineon [49]
$E = 140 \pm 14$ ($v = 0.2 \pm 0.1$)	-260 ± 20	p^+-Poly2, ALP2LV, EM[b] [48]
NA	203 ± 7 203 ± 7	$0.5\,\mu m$, p-doped $10^{16}\,\mathrm{cm}^{-3}$, as deposited and after 30 min annealing at 400 °C, Infineon [49]

a   Abbreviations: deposition temperature (T_{dep}), hydrofluoric (HF) acid and multi-user MEMS process (MUMPS). NA, not applicable.

b   2-μm low-voltage CMOS process ALP2LV from EM Microelectronic-Marin (EM).

Table 3.8 Fracture strength μ and Weibull modulus m obtained from the bulge testing of different polysilicon thin films. The number of membranes measured is also presented.

μ (GPa)	m (No. of membranes)	Deposition conditions and remarks[a]
3.00 and 1.72[b]	22.8 (53) and 20.0 (57)	LPCVD (625 °C), 0.53 and 1.91 μm, annealed at 1050 °C, 1:10, 0.4–0.8 mm wide diaphragms, automated setup [47]
Fracture strains from 0.043 to 0.101%	NA (15 per etchant condition)	LPCVD (630 °C), 4 μm, annealed at 1000 °C, 3 × 3 mm² membranes, integrity studies from exposure to HF solutions [50]

a   Abbreviations: low-pressure chemical vapor deposition (LPCVD) and hydrofluoric (HF) acid. NA, not applicable.
b   Data normalized to an equivalent tensile reference length $L_E = 10^4\,\mu$m. Thicker films are weaker: $\mu = 2.17$ and 1.38 GPa for $t = 0.53$ and 1.91 μm, respectively, for volume fits with $V_E = 32 \times 10^3\,\mu$m³.

rates, doping, deposition temperature and annealing conditions. The fracture characteristics of polysilicon films have been extensively studied from microtensile tests [47]. In contrast, very few results have been reported concerning bulge fracture results, as evidenced by the limited size of Table 3.8. The wafer scale bulge measurements conducted by the authors show that the strength of polysilicon decreases with film thickness, in agreement with microtensile measurements conducted by the authors on the same films and tensile testing conducted by other groups [47, 82]. Weibull moduli up to 23, obtained from bulge tests, have been reported [47].

3.5.4
Metals, Polymers, and Other Materials

In view of their wide application in microsystems and semiconductor processing, the available mechanical properties obtained using the bulge testing of metals, polymers and other materials are listed in Table 3.9.

3.6
Conclusions

This chapter has demonstrated that a large body of knowledge has been accumulated concerning the load deflection of thin films under differential pressure. Depending on their in-plane and out-of-plane dimensions, such structures can be modelled as membranes or thin plates. Boundary conditions governing their response are those given by either simply supported or elastically or rigidly clamped

Table 3.9 Elastic modulus E or plane-strain modulus $E_{ps} = E/(1-v^2)$ and prestress σ_0 obtained from the bulge testing of metals, polymers and other materials obtained by various methods and under different conditions.

E, E_{ps} (GPa)	σ_0 (MPa)	Material, deposition conditions, remarks[a]
E_{ps} = 92	110	Au, PVD (evaporation), 1.8 μm thick, long membranes [64]
NA	100	Au, PVD (evaporation), 26–279 nm, poly- and c-films, 1-mm disk and 3 × 15 mm² diaphragms, rupture at 120–500 MPa [1]
NA	10–100	Ag, PVD (evaporation), polycrystalline films, 1-mm disk and 3 × 15 mm² diaphragms, rupture at 200–500 MPa [1]
E = 69	232 ± 24 167 ± 17	Al, PVD (sputtering), 0.8 μm, as deposited and after 30 min annealing at 400 °C, Infineon [49]
E_{ps} = 70 ± 4	160 and 50	$Al_{0.99}Cu_{0.01}$ alloy, PVD (sputtering), 0.5 μm thick films, 4 × 16 mm² diaphragms, before and after load/unload cycling [65]
E = 130–156	100	Cu, electroplating, 0.89–3 μm, 2.4 × 10 mm² membranes, 0.8% max. strain, 0.05% yield stress from 225 to 300 MPa [66]
260, 330 ± 24 and 220[c]	NA	Cu, $Cu_{0.5}Ni_{0.5}$ and $Cu_{0.7}Ni_{0.3}$, PVD (evaporation), 1–2 μm [2]
E_{ps} = 135	25 to 71	NiFe permalloy, electroplating, 5.5 μm, long diaphragms [67]
E = 50 ± 10 (v = 0.3 ± 0.1)	73 ± 10	Metal 1, ALP2LV, EM [48][b]
E = 55 ± 10 (v = 0.2 ± 0.2)	30 ± 10	Metal 2, ALP2LV, EM [48][b]
E_{ps} = 402 ± 20	125	W, PVD (sputtering), 0.38 μm thick films, 4 × 16 mm² single-layer diaphragms [65]
E = 311	300	Mo, PVD (sputtering), 0.8 μm, 870 × 870 μm² membranes [28]
E = 3.13, 3.22	32.2, 35.2	Polyimide, 5.2–11.4 μm, spin cast, DuPont PI2525 and Hitachi PIQ13, 4.8 mm square and 13 mm circular membranes [2]
E = 3 ± 0.2	30.6 ± 0.9	Polyimide, spin cast, 6–10 μm thickness range, long membranes, 2–10 μm width range, 4% ultimate strain [10]
E = 0.71–8.6	28.7 to 49.4	Polyimide, spin cast, 5–11 μm thickness, square geometry [11]
E = 4.31	21.7 ± 2.1	SU-8 resist, spin cast, 2.25 μm, rectangular long geometry, creep studies, transient compliance D_n of 0.294 GPa^{-1} [70]
E = 3.7 ± 0.2 and 2.0 ± 0.4	35.2 ± 0.2 and 15.8 ± 0.2	PAE polymer, spin cast, 4.6–11.8 μm, dense and 40% porous films, square membranes 1.6 × 1.6 mm² [71]
E_{ps} = 0.00225	0.35	MAA-PEG7, 3-co-MAABPE5% (biocomp. polymer), spin cast, 0.5–1 μm thick, long, single layer membranes [72]
E = 12.7 ± 0.5 and 8.4 ± 0.5	NA	OSG, LPCVD (OMCTS, O_2 and TMCTS, O_2 precursors), 0.5–6 μm, composite Si_3N_4/OSG/SiNx membranes [73]
E = 435 ± 32	323 ± 35	3C-SiC, CVD, 2 μm, square membranes, 1 to 4 mm wide [68]
E = 350–363	120–254	Crystalline SiC, APCVD (1330 °C), 1.6–2.7 μm, 1 mm wide square membranes, fracture μ up to 4.2 GPa, m = 23.5 [58]
E = 308 (v = 0.16)	75	Poly-SiC, APCVD (1050 °C), 2.9 μm, 1 mm wide square membranes, μ = 5.7 GPa, m = 7.8, 24 devices measured [58]
340–420[c]	−1170 and 70	Amorphous SiC, PECVD (200–400 °C), 2–3 μm, 50 mm-diameter circular membranes, before/after anneal, μ = 0.3 GPa [69]
E = 310	82.9	ZnO, PVD (sputtering), 0.4–1 μm, 4 × 4 and 2 × 8 mm² composite Si_xN_y/ZnO membranes [56]
E_{ps} = 114.6 ± 5.9	135 (edge) to −16 (center of wafer)	ZnO, PVD (sputtering), 0.7 μm, long membranes with shorter edge lengths from 0.6 to 1 mm, stacked with tensile SiN_x [59]

a Materials and abbreviations: silver (Ag), gold (Au), aluminum (Al), copper (Cu), nickel (Ni), iron (Fe), tungsten (W), molybdenum (Mo), poly(aryl ethers) (PAE) polymeric low dielectric contant thin film, organosilicate glass (OSG) based dielectric, octamethylcyclotetrasiloxane (OMCTS), tetramethylcyclotetrasiloxane (TMCTS), silicon carbide (SiC), zinc oxide (ZnO), and physical vapor deposition (PVD), chemical vapor deposition (CVD), plasma-enhanced chemical vapor deposition (PECVD), atmospheric pressure chemical vapor deposition (APCVD) and low-pressure chemical vapor deposition (LPCVD). NA, not applicable.

b 2-μm low-voltage CMOS process ALP2LV from EM Microelectronic-Marin (EM).

c Biaxial Young's modulus, $E/(1-v)$.

edges. For square membranes a thorough understanding of the diaphragm mechanics even into the strongly compressive regime permits the reliable extraction of mechanical thin film properties. The second successful paradigm is provided by long membranes, where much of the response occurs in the plane-strain regime. Just as in the case of square membranes, the post-buckling of such structures and its connection to material properties are reasonably well understood.

Experimentally, we are now at a stage where measurements can be performed at the wafer scale in a "quasi-industrial" fashion, thus enabling significant sets of data to be accumulated in a short time. In the authors' opinion, the statistical aspect enabled by these new methods has not been given sufficient attention in the past.

The bulge test also held in reserve the pleasant surprise that fracture mechanical data, as it seems, can be extracted with high reliability. The fact that extremely high Weibull moduli in the range $m = 60$ have been measured on LPCVD silicon nitride and that in silicon oxides, much lower m values and widely differing μ values are observed is a strong indication of the objective value of fracture mechanical parameter determination by the bulge test.

Questions remain open concerning the mechanics of elastically clamped multilayer diaphragms that are not in a state of plane-strain deformation. Such structures are expected and observed to show ripple and meandering transitions like compressively prestressed single-layer diaphragms. How the mechanical properties of the component layers influence these transitions and the response of the structures to applied pressure has not been elucidated to date, whereas a thorough understanding has been obtained in the single-layer case.

The majority of bulge test data have been determined for brittle materials. The more systematic application of this method to metal and organic layers or sandwiches including these layers is expected to raise new questions, related to the possibility of plastic deformation in the materials.

References

1 J. W. Beams, in: *Structure and Properties of thin Films*, C. A. Neugebaur, J. B. Newkirk, D. A. Vermilyea (eds.), Wiley, New York (1959), p. 183–192.

2 T. Tsakalakos, *Thin Solid Films* **75** (1981) 293–295.

3 M. Sheplak, J. Dugundji, *J. Appl. Mech* **65** (1998) 107–115.

4 R. J. Jaccodine, W. A. Schlegel, *J. Appl. Phys.* **37** (1966) 2429–2434.

5 E. I. Bromley, J. N. Randall, D. C. Flanders, R. W. Mountain, *J. Vac. Sci. Technol. B* **1** (1983) 1364–1366.

6 T. Kramer, Mechanical properties of compressive silicon nitride thin films, PhD Thesis, University of Freiburg, 2003.

7 S. Timoshenko, S. Woinowsky-Krieger, Theory of Plates and Shells, McGraw-Hill, New York, 1987.

8 A. E. Love, Treatise on the Mathematical Theory of Elasticity, 4th ed., Dover, New York, 1944.

9 J. Y. Pan, P. Lin, F. Maseeh, S. D. Senturia, IEEE Solid-State Sens. *Actuators Workshop*, Hilton Head, SC, 1988, p. 84–87.

10 M. G. Allen, M. Mehregany, R. T. Howe, S. D. Senturia, *Appl. Phys. Lett.* **51** (1987) 241–243.

11 M. Mehregany, M. G. Allen, S. D. Senturia, *IEEE Solid-State Sensors Workshop*, 1986, p. 58–61.

12 P. Lin, S. D. Senturia, *Mater. Res. Soc. Symp. Proc.*, **188** (1990) 41–46.

13 J. Y. Pan, A study of suspended-membrane and acoustic techniques for the determination of the mechanical properties of thin polymer films, PhD Thesis, MIT, 1991.

14 O. Tabata, K. Kawahata, S. Sugiyama, H. Inagaki, I. Igarashi, *Tech. Digest 7th Sensor Symposium*, 1988, p. 173–176.

15 O. Tabata, K. Kawahata, S. Sugiyama, I. Igarashi, *Sens. Actuators* **20** (1989) 135–141.

16 O. Tabata, K. Shimaoka, S. Sugiyama, I. Igarashi, *Tech. Digest 8th Sensor Symposium*, 1989, p. 149–152.

17 O. Tabata, S. Sugiyama, M. Takigawa, *Appl. Phys. Lett.* **56** (1990) 1314–1316.

18 K. Shimaoka, O. Tabata, S. Sugiyama, M. Takeuchi, *Tech. Digest 12th Sensor Symposium*, 1994, p. 27–30.

19 W. D. Nix, *Metall. Trans.* **20A** (1989) 2217–2245.

20 M. K. Small, J. J. Vlassak, W. D. Nix, *Mater. Res. Soc. Symp. Proc.* **239** (1991) 13–18.

21 J. J. Vlassak, W. D. Nix, *J. Mater. Res.* **7** (1992) 1553–1563.

22 J. J. Vlassak, W. D. Nix, *J. Mater. Res.* **7** (1992) 3242–3249.

23 D. Maier-Schneider, J. Maibach, E. Obermeier, *J. Microelectromech. Syst.* **4** (1995) 238–241.

24 D. Maier-Schneider, J. Maibach, E. Obermeier, *J. Micromech. Microeng.* **2** (1992) 173–175.

25 D. Maier-Schneider, A. Köprülülü, E. Obermeier, *J. Micromech. Microeng.* **5** (1995) 121.

26 D. Maier-Schneider, A. Ersoy, J. Maibach, D. Schneider, E. Obermeier, *Sens. Mater.* **7** (1995) 121–129.

27 A. Bosseboeuf, M. Boutry, T. Bourouina, *Microsyst. Microanal. Microstruct.* **8** (1997) 261–272.

28 R. Yahiaoui, K. Danaie, S. Petitgrand, A. Bosseboeuf, *Proc. SPIE* **4400** (2001) 160–169.

29 K. Danaie, A. Bosseboeuf, R. Yahiaoui, S. Petitgrand, *Proc. MME 2001*, 2001, p. 297–300.

30 K. Danaie, A. Bosseboeuf, C. Clerc, C. Gousset, G. Julie, *Sens. Actuators A* **8** (2002) 78–81.

31 V. Ziebart, O. Paul, U. Münch, J. Shwizer, H. Baltes, *J. Microelectromech. Syst.* **7** (1998) 320–328.

32 O. Paul, H. Baltes, *J. Micromech. Microeng.* **9** (1999) 19–29.

33 J. Gaspar, P. Ruther, O. Paul, *Mater. Res. Soc. Symp. Proc.* **977E** (2006) FF8.8.

34 J. Yang, O. Paul, *Sens. Actuators A* **97–98** (2002) 520–526.

35 J. Gaspar, P. Ruther, O. Paul, *Proc. 3rd Int. Conf. Multiscale Materials Modeling*, 2006, p. 966–968.

36 V. Ziebart, O. Paul, H. Baltes, *J. Microelectromech. Syst.* **8** (1999) 423–432.

37 O. Paul, T. Kramer, *Dig. Tech. Papers Transducers 2003*, Boston, 2003, p. 432–435.

38 V. Ziebart, O. Paul, U. Münch, H. Baltes, *Mater. Res. Soc. Symp. Proc.* **505** (1998) 27–32.

39 T. Kramer, O. Paul, *J. Micromech. Microeng.* **12** (2002) 475–478.

40 T. Kramer, O. Paul, *Proc. MEMS 2003 Conference*, Kyoto, 2003, p. 678–681.

41 V. Ziebart, O. Paul, H. Baltes, *Mater. Res. Soc. Symp. Proc.* **546** (1999) 103–108.

42 T. Kramer, O. Paul, *Sens. Actuators A* **92** (2001) 292–298.

43 E. Bonnotte, P. Delobelle, L. Bornier, *J. Mater. Res.* **12** (1997) 2234–2248.

44 D. S. Popescu, T. S. J. Lammerink, M. Elwenspoeck, *Proc. MEMS 94*, Oiso, 1994, p. 188–192.

45 R. I. Pratt, G. C. Johnson, *Mater. Res. Soc. Symp. Proc.* **308** (1993) 115–120.

46 S. Jayaraman, R. L. Edwards, K. J. Hemker, *J. Mater. Res. Soc.* **14** (1999) 688–697.

47 J. Gaspar, M. Schmidt, O. Paul, *Dig. Tech. Papers Transducers 2007*, Lyon, 2007, in press.

48 D. Jaeggi, *Thermal converters by CMOS technology*, PhD Thesis, No. 11567, ETH Zurich, 1996.

49 H. Kapels, D. Maier-Schneider, R. Schneider, C. Hierold, *Proc. Eurosensors XIII*, The Hague, 1999, p. 393–396.

50 J. A. Walker, K. J. Gabriel, M. Mehregany, *IEEE Proc. MEMS 1990*, 1990, p. 56–60.

51 N. Sabaté, I. Gràcia, J. Santander, L. Fonseca, E. Figueras, C. Cané, J. R. Morante, *Sens. Actuators A* **125** (2006) 260–266.

52 J. Yang, O. Paul, *unpublished results*.

53 R. J. Jaccodine, W. A. Schlegel, *J. Appl. Phys.* **37** (1966) 2429–2434.

54 D. Sobek, A. M. Young, M. L. Gray, S. D. Senturia, *IEEE MEMS Proc. 1993*, 1993, p. 219–224.

55 X. Y. Ye, J. H. Zhang, T. Y. Zhou, Y. Yang, *IEEE 7th Int. Symp. on Micro Machine and Human Science*, 1996, p. 125–129.

56 R. A. Stewart, J. Kim, E. S. Kim, R. M. White, R. S. Muller, *Sens. Mater.* **2** (1991) 285–298.

57 G. F. Cardinale, R. W. Tustison, *Thin Solid Films* **207** (1992) 126–130.

58 N. N. Nemeth, J. L. Palko, C. A. Zorman, O. Jadaan, J. S. Mitchell, *MEMS: Mechanics and Measurements, Society of Experimental Mechanics Symposium*, Portland, OR, 2001, p. 46–51.

59 S. Koller, V. Ziebart, O. Paul, O. Brand, H. Baltes, P. M. Sarro, M. J. Vellekoop, *Proc. SPIE* **3328** (1998) 102–109.

60 R. L. Edwards, G. Coles, W. N. Sharpe, *Exp. Mech.* **44** (2004) 49–54.

61 J. Gaspar, T. Smorodin, M. Stecher, O. Paul, unpublished results: the films were deposited by Infineon and measured at IMTEK using an automated full wafer bulge setup.

62 T. Kramer, O. Paul, *Mater. Res. Soc. Symp. Proc.* **695** (2002) 79–84.

63 O. R. Shojaei, A. Karimi, *Thin Solid Films* **332** (1998) 202–208.

64 R. P. Vinci, J. J. Vlassak, *Annu. Rev. Mater. Sci* **26** (1996) 431–462.

65 A. J. Kalkman, A. H. Verbruggen, G. C. A. M. Janssen, F. H. Groen, *Rev. Sci. Instrum.* **70** (1999) 4026–4031.

66 Y. Xiang, X. Chen, J. J. Vlassak, *Mater. Res. Soc. Symp. Proc.* **695** (2002) L4.9, 189–194.

67 J. T. Ravnkilde, V. Ziebart, O. Hansen, H. Baltes, *Proc. Eurosensors XIII*, The Hague, 1999, p. 383–386.

68 J. S. Mitchell, C. A. Zorman, T. Kicher, S. Roy, M. Mehran, *J. Aerospace Eng.* **16** (2003) 46–54.

69 H. Windischmann, *J. Vac. Sci. Technol. A* **9** (1991) 2459–2463.

70 B. Schoeberle, M. Wendlandt, C. Hierold, *Proc. Eurosensors XX*, Göteborg, 2006, p. 354–355.

71 D. W. Zheng, Y. H. Xu, Y. P. Tsai, K. N. Tu, P. Patterson, B. Zhao, Q.-Z. Liu, M. Brongo, *Appl. Phys. Lett.* **76** (2000) 2008–2010.

72 J. Gaspar, O. Paul, unpublished results: polymer films prepared by O. Prucker and W. Betz (CPI group, IMTEK) and measured by the authors.

73 Y. Xiang, X. Chen, T. Y. Tsui, J.-I. Jang, J. J. Vlassak, *J. Mater. Res.* **21** (2006) 386–395.

74 Y. Xu, Y. Tsai, D. W. Zheng, K. N. Tu, C. W. Ong, C. L. Choi, B. Zhao, Q. Z. Liu, M. Brongo, *J. Appl. Phys.* **88** (2000) 5744–5750.

75 V. Ziebart, *Mechanical properties of CMOS thin films, PhD Thesis*, No. 13457, ETH Zurich, 1999.

76 L. D. Landau, E. M. Lifshitz, Theory of Elasticity, 3rd ed., Butterworth-Heinemann, London, 1986.

77 G. Gerlach, A. Schroth, P. Pertsch, *Sens. Mater.* **8** (1996) 79–98.

78 S. Levy, in: Non-Linear Problems in Mechanics of Continua, Proc. Symp. in Applied Mathematics (1949) Vol. 1, p. 197–210.

79 J. J. Vlassak, New experimental techniques and analysis methods for the study of the mechanical properties of materials in small volumes, PhD Thesis, Stanford University, Palo Alto, CA, 1994.

80 W. H. Press, S. A. Teukolsky, W. T. Vetterling, B. P. Flannery, Numerical Recipes in C, Cambridge University Press, Cambridge, 1995, p. 420–423.

81 O. Tabata, T. Tsuchiya, N. Fujitsuka, *Tech. Dig. 12th Sensor Symposium*, 1994, p. 19–22.

82 O. M. Jadaan, N. N. Nemeth, J. Bagdahn, W. N. Sharpe, *J. Mater. Sci.* **38** (2003) 4087–4113.

83 W. Weibull, *J. Appl. Mech.* **18** (1951) 293–297.

84 D. G. S. Davies, *Proc. Br. Ceram. Soc.* **22** (1973) 429–452.

85 E. Y. Robinson, *Some Problems in the Estimation and Application of Weibull Statistics*, University of California, UCRL 70555, 1967.

86 C. Liu, in: *Foundations of MEMS*, Pearson Education (Prentice Hall), 2006, Ch. 10.

87 F. E. Rasmussen, B. Geilman, M. Heschel, O. Hansen, A. M. Jorgensen, *Proc. 1st Symp. Sci. Tech. of Dielectrics in Emerging Fields*, Paris, 2003, p. 218–229.

88 J. Gaspar, O. Paul, unpublished results: the authors conducted full wafer bulge measurements of LPCVD Si3N4 grown at 770 °C and subsequently annealed at temperatures up to 1050 °C. The plane-strain modulus, prestress and fracture strength vary as little as 1.1%, 1.5% and 1.4% per 100 °C, respectively, for temperatures between 650 and 1050 °C.

89 A. Stoffel, A. Kovács, W. Kronast, B. Müller, *J. Micromech. Microeng.* **6** (1996) 1–13.

90 J. G. E. Gardeniers, H. A. C. Tilmans, C. C. G. Visser, *J. Vac. Sci. Technol. A* **12** (1996) 2879–2892.

91 S. S. Verbridge, J. M. Parpia, R. B. Reichenbach, L. M. Bellan, H. G. Craighead, *J. Appl. Phys.* **99** (2006) 124304–124311.

92 A. Kaushik, H. Kahn, A. H. Heuer, *J. Microelectromech. Syst.* **12** (2005) 359–367.

93 L. U. J. T. Ogbuji, D. R. Harding, *Thin Solid Films* **263** (1995) 194–197.

94 M. Stadtmüeller, *J. Electrochem. Soc.* **139** (1992) 3669–3674.

95 G. Smolinsky, T. P. H. F. Wendling, *J. Electrochem. Soc.* **132** (1985) 950–954.

96 M. L. Hammond, *Mater. Res. Soc. Symp. Proc.* **182** (1990) 3.

97 R. S. Roller, *Solid State Technol.* **20** (1977) 63.

98 Y. Wang, R. Ballarini, H. Khan, A. H. Heuer, *J. Microelectromech. Syst.* **14** (2005) 160–166.

4
Uniaxial Tensile Test for MEMS Materials

Takahiro Namazu, Department of Mechanical and Systems Engineering, University of Hyogo, Japan

Abstract
Investigation of the mechanical behavior in materials with micro- or nano-scale dimensions is one of the significant concerns in the MEMS field because the knowledge that has been obtained in the material tests is directly linked to the performance and reliability of MEMS devices. For such the minute specimens, the uniaxial tensile test is the most popular material evaluation method as with the macroscopic specimens. The many efforts to accurately measure the mechanical properties through the test are being made consistently. This chapter is organized basically by three topics. First, the uniaxial tensile test techniques for a film specimen are described. In particular, the technical issues regarding specimen chucking and strain measurement are mentioned. Second, the material constants that should be estimated for the MEMS design are presented. In this topic, the four different materials, such as a brittle material, a metallic material, a polymer film and a shape memory alloy film, have been covered. Third, a short description about the fatigue test is given. Finally, a brief outlook towards the development of high-quality MEMS devices is stated in the light of the importance of the reliable design that is conducted using the appropriate material constants.

Keywords
MEMS; tensile test; films; specimen chucking; strain measurement; material constant; fatigue; reliable design

Reliability of MEMS.
Edited by O. Tabata, T. Tsuchiya
Copyright © 2013 WILEY-VCH Verlag GmbH & Co. KGaA, Weinheim
ISBN: 978-3-527-33501-5

4.1
Introduction

The uniaxial tensile test is the most common and variable method for the mechanical characterization of materials. The tensile test measures the resistance of a material to a static or slowly applied tensile force. The result of the tensile test is commonly displayed in a stress–strain diagram, providing fundamental material information concerning Young's modulus, Poisson's ratio, fracture strength and ductility. These are, needless to say, very important material parameters that are indispensable for use in the structural design of mechanical components. In the field of microelectromechanical systems (MEMS), the uniaxial tensile test is very popular for characterizing the mechanical behavior of a specimen with micro/nanoscale dimensions. To date, many researchers have made efforts to develop a variety of original tensile test techniques for the accurate mechanical characterization of micro/nanoscale materials used in the construction of MEMS.

In many of today's MEMS technologies, one primary emphasis is on the accurate measurement of the mechanical properties of MEMS materials and integration of the measured properties into the design of MEMS devices. In this chapter, the uniaxial tensile test techniques intended for micro/nanoscale specimens fabricated by micromachining technologies are described first. In particular, focus is concentrated on several technical issues related to specimen chucking and strain measurement and a unique test method for investigating nanoscale specimens is introduced. Second, the evaluation of material properties for several types of representative MEMS materials, such as brittle and metallic materials, polymer films and shape memory alloy films, are described. The importance of material property measurements at the micro/nanoscale is mentioned in the light of the requirement for reliable structural design of MEMS devices. Finally, a common technical description regarding the fatigue test of a film specimen is presented, along with experimental results.

4.2
Technical Issues for Uniaxial Tensile Testing of Film Specimens

In the last few decades, many experiments have been performed in order to characterize microscale MEMS materials in terms of their fundamental mechanical properties. Mechanical tests, such as the tensile test, bending test, indentation test and bulge test, are the most common methods used to measure the material properties of a microscale specimen [1].

The bending test, which measures the resistance of a beam-shaped material to a slowly applied bending force, has some advantages, such as the ease of measuring the bending force and deflection and the simplicity of the test set-up. Specimen preparation is also relatively simple, because the specimen shape is not very complicated. However, when considering the dependence of specimen size on the bending strength, much attention should be given, because the bending motion generates a stress gradient in a specimen [2]. In the case of the cantilever bending test, it must also be considered that movement of the loading point occurs with increase in the bending force [3].

The bulge test, which measures the deflection of a membrane under a static pressure, can easily be used to evaluate the mechanical properties of a thin film. However, there are the several technical problems, for instance, difficulty in measuring accurately the Young's modulus or the fracture strength, due to a stress concentration at the edge of the membrane, and the membrane shape dependence on the measured property.

The indentation test is one of the most traditional test methods for characterizing thin-film specimens. The film specimen preparation and the test set-up are very simple; however, both the elasticity of the substrate and the tip curvature of the indenter affect the estimation of film properties in the case of a very thin film and a relatively large penetration depth [4]. Additionally, the Poisson's ratio of a film specimen has to be assumed when the Young's modulus is determined from the outcome of an indentation test.

The tensile test measures the deformation response of a material produced by applying a tensile force to a specimen. The strong point of the tensile test is that it is possible to apply a uniform stress distribution to a specimen, and this is advantageous for precision in the investigation of material characteristics. The test directly yields a stress–strain diagram, which is integral to the understanding of various aspects of the material under examination [5]. A schematic illustration of a tensile test set-up is shown in Figure 4.1. A film specimen is prepared using conventional micromachining processes, including photolithography, etching and deposition. The specimen placed in the tensile test apparatus is tensioned using a type of actuator, such as a piezoelectric actuator, that provides a smooth displacement at a constant strain rate. Extension of the specimen is measured using a type of high-resolution displacement meter when the tensile force is applied.

Figure 4.2 depicts schematically a typical tensile stress–strain diagram illustrating the many important characteristics that are evident. During the initial stress period, the linear elongation of the material in response to the tensile stress is known as elastic deformation. The slope in the elastic region provides the modulus of elasticity (Young's modulus, *E*), expressed using Hooke's law:

$$E = \frac{d\sigma}{d\varepsilon} \tag{1}$$

where $d\sigma$ and $d\varepsilon$ are the applied stress increment and the strain increment in the elastic region, respectively. The Young's modulus is a significant material

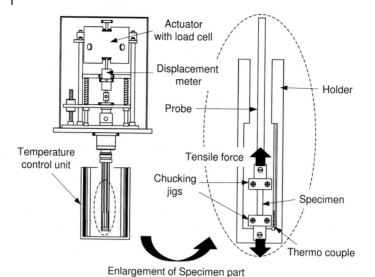

Enlargement of Specimen part

Figure 4.1 Schematic of a uniaxial tensile test set-up for a film specimen. The tensile tester typically includes an actuator, load cell, displacement meter, specimen holder and temperature control unit.

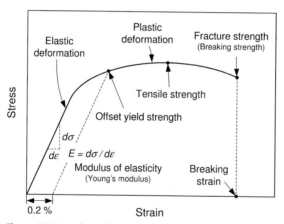

Figure 4.2 Typical tensile stress–strain diagram for a ductile material. The stress–strain diagram shows many important material characteristics of a test specimen.

parameter that reveals how stiff or rigid a material is. In other words, it expresses the resistance of a material to elastic elongation. For a ductile material, when the stress exceeds a critical stress value (yield strength, σ_y), some of the deformation of the material becomes permanent and this is termed plastic deformation. The yield point separates the elastic region from the total deformation range of a material. The maximum stress that a material can sustain is called the tensile strength,

σ_T. At this point, necking starts to occur, signified by a local decrease in the cross-sectional area. The fracture strength, σ_f, is the strength at the breaking point. Furthermore, if a lateral strain is measured during the tensile test, the Poisson's ratio, v, can be determined. The Poisson's ratio, which is represented as the ratio of the lateral elastic strain, $\varepsilon_{lateral}$, to the longitudinal strain, $\varepsilon_{longitudinal}$, is given by

$$v = -\frac{\varepsilon_{lateral}}{\varepsilon_{longitudinal}} \tag{2}$$

The Poisson's ratio for many metals is typically 0.3 in the elastic region. In the plastic region, the ratio increases to approximately 0.5, because the volume is constant during plastic deformation.

In the MEMS field, however, because of the dimensionally shrunk specimens that are targeted, there are several technical problems that need to be overcome in order to perform a tensile test such the specimens. These include challenges in the design and manufacture of both of an original tensile test specimen and tensile tester and the difficulties involved in properly fastening a specimen and in accurately measuring minute tensile force and strain. If researchers can find solutions to these problems, the tensile test will become a promising method for the accurate characterization of micro/nanoscale MEMS materials. In the sections that follow, three recent and significant technologies are introduced regarding uniaxial tensile testing of micro/nanoscale materials: specimen chucking, strain measurement and characterization of nanoscale specimens. The technical problems that should be solved are considered.

4.2.1
Specimen Chucking

The major technical issues in the uniaxial tensile testing of film specimens are commonly regarded as fastening of miniature specimens and precision alignment of the specimen axial direction with the tensile direction. These problems are of significant concern for the tensile testing of bulk specimens, but they become even more important technical factors in acquiring reasonable results from experiments with miniature specimens.

Many researchers have been challenged and have developed novel technologies for the fixing of film specimens. Chucking methods that have often been applied recently are broadly classified into three categories: electrostatic, mechanical and adhesion chucking.

Electrostatic chucking is a method that uses an electrostatic force to hold a film specimen. This method is advantageous for operation of the test and reproducibility of the fixing position of a specimen. Tsuchiya et al. [6] fabricated a film specimen by surface micromachining and found that an attractive electrostatic force could be used to hold the sample during tensile loading. As shown in Figure 4.3, a large plate, called a paddle, is laid on the free end of the specimen beam, while the other end is fixed to a substrate. After alignment, the probe grabs

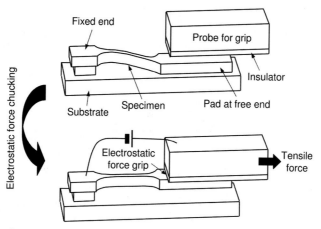

Figure 4.3 Specimen chucking method by means of an electrostatic force grip.

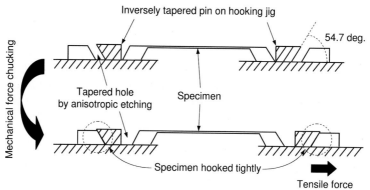

Figure 4.4 Schematic of mechanical hooking of a film specimen. Tapered holes etched in the specimen grips are hooked mechanically by inversely tapered pins.

the paddle using the electrostatic force generated by the difference in potential between the sample and the probe. By moving the sample stage against a fixed probe, a tensile force is then applied to the specimen until failure occurs. After the specimen has fractured, the fractured end is released by application of an opposite-polarity voltage between the probe and the substrate.

Mechanical chucking is the most orthodox method used to fix a specimen. This method has several styles, for example, pin-hole hooking, clipping and grabbing. Isono et al. [7] developed a mechanical hooking method, shown in Figure 4.4, which enables a specimen with a wet-etched tapered hole of 54.7° to be mechanically hooked using an inversely tapered pin with the same angles. The rigidity of chucking improves with increase in the tensile force, because the tapered angle causes a force that pushes the specimen towards the specimen holder. Namazu et al. [8] employed a simpler hooking method that includes a simple-shape pin and hole. Figure 4.5 show a photograph of a micromachined film specimen and

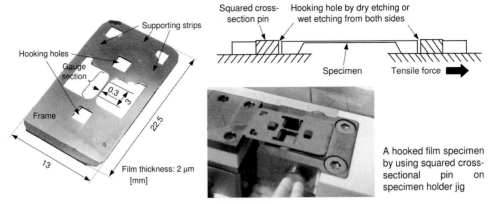

Figure 4.5 Photograph of a micromachined film specimen and a schematic of typical pin-hole chucking. Conventional micromachining processes, such as photolithography, film deposition and etching, can produce a precision film specimen for the uniaxial tensile test.

a schematic of pin-hole hooking. A film specimen was fabricated and a gauge section including a parallel part for testing, a frame section with square hooking holes and a supporting strip section were micromachined into the specimen. The specimen was then fastened to the tensile tester by hooking a pin with a square or circular cross-section into the hole fabricated in the specimen. Such mechanical chucking methods are effective for fixing a relatively larger specimen with a thickness of tens of microns, but these methods are somewhat disadvantageous for smaller specimens with a thickness of less than 1 μm. This is because in fastening the specimen, a small gap between the pin and the hole easily imparts a torsional or flexural irregular deformation to the specimen, which can possibly lead to unexpected failure of the specimen.

The adhesion-chucking method uses an adhesive or glue and excels in ease of fixing a specimen. This method does not require a complicated test set-up, but the testing atmosphere is restricted to room temperature, because of the low thermal resistance of most adhesives. In addition, the adhesion bond usually has a lower stiffness than most representative MEMS material specimens, so that the bond may deform easily when a large tensile force is applied to the specimen.

All of the chucking methods described above have their respective technical merits in fixing a film specimen. However, when a specimen is set on a tester, a particular skill for high-precision positioning is required to align the specimen's longitudinal direction with the direction of the load. To solve this problem, an on-chip tensile test method was proposed, which allows ease of handling and alignment of a miniature specimen. Sato and Shikida [9] developed the alignmentless uniaxial tensile test device shown in Figure 4.6. The device includes a film specimen, torsion beam and loading lever sections within a diced square Si chip ~150 mm² in size. The specimen is tensioned via a torsion beam by application of a normal force to the loading lever. This type of tensile test does not require a specimen holding mechanism and all the parts can be precisely formed at the same

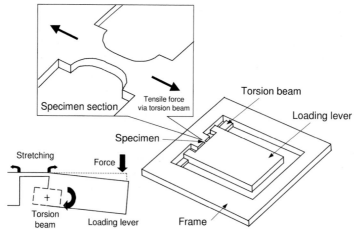

Figure 4.6 On-chip film tensile testing device including film specimen, loading lever, torsion beam and frame. When applying a force to the loading lever, the specimen is uniformly tensioned via a torsion beam without alignment between specimen and the direction of loading.

time through conventional micromachining technologies. The on-chip tensile test technique, without the normally crucial chucking, is effective for the evaluation of the tensile properties of micron/submicron dimensional specimens.

4.2.2
Strain Measurement

In addition to specimen chucking, a direct strain measurement during tensile testing is a significant concern, due to less deformation in a film specimen than in a macro specimen. The fine strain measurement of miniature specimens provides results allowing the accurate determination of material characteristics.

To date, direct strain measurement methods have been developed that are categorized by the type of measurement technique used, such as optical microscopy [10], laser interferometry [11], atomic force microscopy (AFM) [7, 12], scanning electron microscopy (SEM) [13, 14] and X-ray diffraction (XRD) [8, 15, 16]. Optical microscopy provides several techniques for the measurement of specimen deformation. A single or dual field-of-view optical microscope can also be employed to measure the deformation of a specimen. The measurement resolution depends on the pixel quality of the digital camera and the magnification of the microscope. An image data processing system also plays a key role in the resolution. For example, image processing using a template-matching algorithm allows the measurement of displacement with accuracy to the sub-pixel level, from the image data obtained before and after deformation of a specimen.

A laser interferometric measurement system provides non-contact displacement measurement as with the optical microscope method. The interferometric method is able to measure transverse displacement in addition to axial displacement during tensile testing. Irradiation of a laser beam to a fabricated line pattern on the gauge section of a specimen produces interference fringes. Photodiode sensors detect changes in the fringe pattern, permitting the determination of biaxial planar deformations. The advantage of this method is achieving non-contact measurement of deformation in a very small specimen without a large time lag.

The AFM system, including a three-axis piezoelectric or voice-coil actuation scanner and a laser deflection feedback system, has a high-resolution positioning control function that can precisely obtain a topographical image of a specimen surface with nanoscale irregularities. Biaxial planar deformations of a film specimen can be measured with a uniaxial tensile tester equipped with an AFM observation function. AFM tensile testing allows axial and transverse deformations to be measured with high resolution by scanning the line patterns on the specimen using an AFM probe (see Figure 4.7). The AFM method has succeeded in the determination of not only the Young's modulus, but also the Poisson's ratio of typical MEMS materials, such as micron- or submicron-thick single/polycrystalline Si and diamond-like carbon (DLC) films prepared by low-pressure chemical vapor deposition (CVD).

The SEM observation method allows not only displacement measurements, but also a microstructural experimental approach with *in situ* observations of specimen deformation. It is often the case that the SEM method focuses on the characterization of deformation behavior and observation of microcrack propagation in film specimens, from the technical point of view that electron beam scanning enables the image of a specimen to be captured.

XRD, which is commonly used for crystallographic structural analysis, can be employed to measure elastic strain in a film specimen. Figure 4.8 shows a photograph of a tensile tester set into an XRD apparatus. Correct setting of the tester

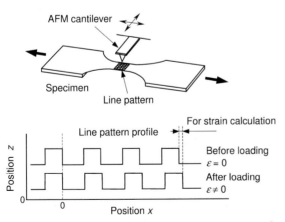

Figure 4.7 AFM tensile testing for the direct measurement of elongation on the longitudinal and transverse axes. As the tensile force increases, line patterns on the specimen deform. The deformation is accurately measured with AFM.

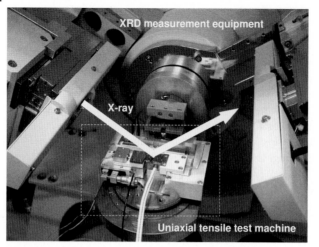

Figure 4.8 Photograph of uniaxial tensile tester equipped with XRD apparatus. A compact tensile tester is set on the sample stage.

on the sample stage of the XRD apparatus requires optimum design to achieve a smaller size tester. The tensile tester, measuring $130 \times 80 \times 30$ mm, consists of a piezoelectric actuator built into the actuator case, a linear variable differential transformer (LVDT), a load cell and specimen holders. The compact tensile tester allows the measurement of out-of-plane elastic strain, which is calculated from a shift in the XRD peak angle with an alteration of tensile stress. If the scattering vector of an X-ray is normal to a specimen, the lattice spacing, d, of the lattice plane parallel to the specimen surface, as calculated from Bragg's law, is given by

$$d = \frac{\lambda}{2\sin\theta} \tag{3}$$

where λ is the wavelength of the X-ray and θ is the diffraction angle.

Figure 4.9 illustrates the concept of the XRD tensile test. As the tensile force increases during the course of testing, the lattice spacing decreases, causing the XRD peak position to shift towards a higher angle. Measuring the peak angle for a certain interval of tensile force permits the determination of out-of-plane elastic strain with atomic resolution.

Figure 4.10 shows a representative example of an XRD tensile test result for a single-crystal Si specimen. XRD measurements were carried out at various tensile stresses. The axial direction of the specimen is oriented with the [110] direction in the (001) plane. The XRD peak angle at the lowest stress is ~69.2°, a typical diffraction angle for Si (004). The trend of the XRD peak shift to a higher angle can be observed as the tensile stress increases. The XRD tensile test method can

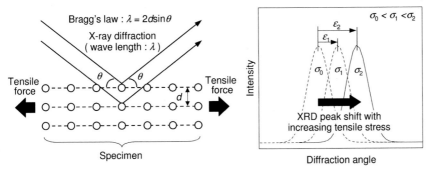

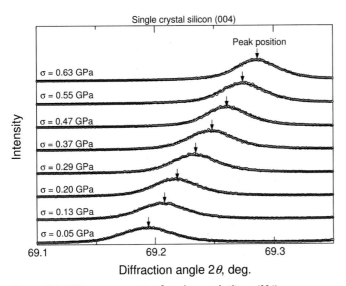

Figure 4.9 Schematic for out-of-plane strain determination in XRD tensile testing. As tensile force increases, the lattice spacing in the out-of-plane direction of a specimen decreases by just the Poisson's ratio. The shift in the lattice spacing, which corresponds to an out-of-plane strain, can be measured by means of XRD and leads to the determination of the Poisson's ratio for the crystalline film specimen.

Figure 4.10 XRD measurements of single-crystal silicon (004) at various tensile stresses. An increase in tensile stress causes a shift in the XRD peak position towards higher angle.

quantify the out-of-plane Poisson's ratio for a film specimen, although the materials under evaluation are confined to those with crystalline structure.

When a film material, utilized as a test piece, is fragile due to high internal stresses, then a laminated specimen consisting of the film deposited on a substrate may be used for the XRD tensile test. In the case of a laminated specimen, the effective elastic modulus is the composite elastic modulus of the film and substrate and is estimated from the slope of the stress–elastic strain relationship. Based on the assumption that the axial strain of a film is equal to that of the substrate, the following equation gives the effective elastic modulus, E_{1+2}, of the composite specimen:

$$E_{1+2} = \frac{t_1}{t_1+t_2} \frac{E_1}{1-v_1^2} + \frac{t_2}{t_1+t_2} \frac{E_2}{1-v_2^2} - \frac{\left(\dfrac{t_1}{t_1+t_2}\dfrac{E_1 v_1}{1-v_1^2} + \dfrac{t_2}{t_1+t_2}\dfrac{E_2 v_2}{1-v_2^2}\right)^2}{\dfrac{t_1}{t_1+t_2}\dfrac{E_1}{1-v_1^2} + \dfrac{t_2}{t_1+t_2}\dfrac{E_2}{1-v_2^2}} \tag{4}$$

where E_1, v_1 and t_1 are the Young's modulus, the in-plane Poisson's ratio and the thickness of the substrate, respectively, and E_2, v_2 and t_2 are the same parameters for the film. Assuming that the out-of-plane Poisson's ratio is the same as the in-plane ratio, E_2 can be estimated from Eq. (4) if E_{1+2}, E_1, v_1 and v_2 are precisely measured from the XRD tensile tests. The XRD tensile test method is also applied to the observation of transitions in crystallographic structures. For example, in the case of shape memory alloy thin films, the phase transformation between martensite and austenite around the phase transition temperature can be observed from the change in the tensile stress or temperature during the test.

Accurate strain measurement during tensile testing is required in order to obtain representative material characteristics of specimens at a micro- or nanolevel. Additionally, the measurement of specimen dimensions influences the material property calculations. When measuring dimensions before or after testing, several types of microscopes are usually employed, such as SEM, AFM, laser microscope and optical microscope. The determination of absolute dimensions in a micro/nanoscale specimen involves many technical difficulties. Therefore, cross-comparison among these microscopes should be performed for confirmation of measurement.

4.2.3
Tensile Testing of Nanoscale Structures

In a tensile test set-up, specimen dimension shrinkage towards the submicronor nanoscale produces more technical difficulties, such as specimen chucking, loading and measurement of both tensile force and displacement. To overcome these issues, several proposals have been made for the tensile testing of specimens with nanoscale dimensions. Kiuchi et al. [17] developed an on-chip tensile test device that can examine Young's modulus and the fracture strength of an amorphous carbon nanowire deposited by a focused ion beam-assisted CVD method. The device integrates an electrostatic force-driven comb actuator and a displacement measurement lever in an Si chip. The lever amplifies the specimen elongation by a factor of 91 and the deflection of the lever that indicates the amplified specimen elongation can be estimated by image processing using a CCD camera. The tensile force applied to a specimen is measured by the difference in the lever movement before and after failure. Ishida et al. [14] observed adhesion phenomena of single-crystal Si nanoprobes using transmission electron microscopy. Two Si nanoprobes that faced one another and were fabricated with a tip radius of a few nanometers could be moved by an electrothermal force-driven actuator in order to contact or detach from each other. The process of Si atom rearrangement was clearly observed when one probe tip detached from the other.

The on-chip tensile test device, which consists of a nanoscale specimen, actuator and a force/displacement measurement mechanism within a single Si chip, can be precisely fabricated through conventional MEMS technologies. On-chip tensile testing has the possibility of becoming a standard method for the investigation of the mechanical behavior in materials with submicron- or nanometer-scale dimensions.

4.3
Quantification of Tensile Mechanical Characteristics for Structural Design of MEMS

Many of the micro/nanoscale components in MEMS devices are made from a wide variety of materials that include metals, polymers, ceramics and several types of functional films. The construction materials used in MEMS are not limited to Si and related materials, although they are the most commonly used materials. Several properties and the behavior of materials at a micro- or nano-level are known to differ greatly from those at a macro-level [18]. Understanding of the micro/nanostructures in materials has received considerable attention. Quantifying the mechanical characteristics of materials within a scale that is used practically in MEMS is critical to the design of MEMS. In addition, in the case of film materials prepared by CVD or physical vapor deposition (PVD) methods, such as sputtering, the material properties are dependent on the deposition conditions, so that the influence of the deposition conditions should be examined. Application of specific material properties to a MEMS design performed by computer simulation, such as finite element analysis, will lead to the development of high-performance and reliable MEMS.

In this section, we describe the material property determination method intended for use with structural MEMS materials, including brittle materials, metallic materials, polymer films and shape memory alloy (SMA) films, and quantification of a material's constitutive relationship is covered in detail. We also provide descriptions of some experimental results that have been obtained from uniaxial tensile testing of such materials.

4.3.1
Brittle Materials

Many MEMS devices consist of brittle materials, such as single/polycrystalline Si, silicon nitride (SiN_x) and silicon oxide (SiO_x). These are the most commonly used materials in MEMS devices, because MEMS technology originated from the semiconductor industry. Single crystal Si, which is well known to fail in a brittle manner at ambient temperature, is employed as a structural material in MEMS by virtue of its high Young's modulus, high hardness and excellent formability [19]. To conduct a structural design of an Si MEMS that usually operates at intermediate temperatures (~573 K at the highest), mechanical properties such as the Young's modulus, Poisson's ratio and strength should be evaluated at temperatures from room temperature to 573 K.

Figure 4.11 shows a typical tensile stress–strain diagram of a microscale single-crystal Si specimen. The solid lines indicate the axial strain measured by a linear variable differential transformer (LVDT) and the plotted points are indicative of out-of-plane strain measured by XRD. The Si specimens, oriented with the [110] direction in the (001) plane, show a linear stress–strain relationship indicating brittle failure below 573 K with almost no plastic deformation [15, 16]. In other words, all the specimens have failed before the yield strength has been reached. At room temperature, the Young's modulus and the out-of-plane Poisson's ratio of the specimen are found to be 169 GPa and 0.35, respectively, which are in very close agreement with analytical values [20]. The Young's modulus gradually decreases with elevation of the temperature, whereas the Poisson's ratio does not change. These elastic properties do not depend on the specimen size, which ranges from millimeter to sub-micrometer, whereas the fracture strength does vary considerably with change in size. The strength of Si is also known to have a significant dependence on the surface roughness and the loading method (tension or flexure). Therefore, with regard to the strength, a statistical approach is needed.

Comprehension of important factors affecting the fracture strength is essential when evaluating the strength. The size, distribution and the number of intrinsic

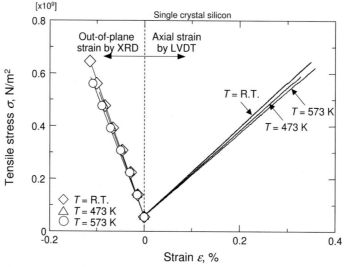

Figure 4.11 Typical tensile stress–strain diagram for single-crystal silicon. The plain lines and the lines with symbols indicate axial and out-of-plane strains measured by LVDT and XRD, respectively.

flaws in a specimen are considered as the main dominating factors in the strength of brittle materials such as Si. Weibull statistics are used to describe and characterize the variability in strength. The strength is given by the Weibull distribution, which is an indicator of the variability of strength resulting from a distribution of flaw sizes. The three-parameter Weibull distribution function, derived from the weakest link theory, is expressed as

$$F(x) = 1 - \exp\left[-\frac{(x-\gamma)^m}{\alpha}\right], \quad x > \gamma \tag{5}$$

where $F(x)$ is the fracture probability and α, m and γ are the Weibull parameters. Taking the derivative of Eq. (5) with respect to x gives the probability density function:

$$f(x) = \frac{m}{\alpha}(x-\gamma)^{m-1}\exp\left[-\frac{(x-\gamma)^m}{\alpha}\right], \quad x \geq \gamma \tag{6}$$
$$= 0 \qquad\qquad\qquad\qquad\qquad x < \gamma$$

The meaning of the Weibull parameters is clarified if Eq. (6) can be rewritten as follows:

$$f(x)dx = \exp\left[-\left(\frac{x-\gamma}{\alpha^{1/m}}\right)^m\right]d\left(\frac{x-\gamma}{\alpha^{1/m}}\right)^m, \quad x \geq \gamma \tag{7}$$
$$= \exp(-Z^m)d(Z^m), \qquad\qquad Z = \frac{x-\gamma}{\alpha^{1/m}} \geq 0$$

In the exponent, Z, $\alpha^{1/m}$ is a unit scale of x, so that α is referred to as a scale parameter. In the case of $m > 1$, $m = 1$ or $m < 1$, $f(x)$ exhibits a different shape for each condition, so that m is termed a shape parameter. γ is called a location parameter, because γ determines the sign of $f(x)$ that takes a positive value in $x \geq \gamma$. In conducting statistical calculations with regard to the strength of materials, the meanings of these parameters are defined as follows: x is the applied stress, α is the representative characteristic strength corresponding to the strength at a fracture probability of 63.2%, m is the Weibull modulus indicating the strength variability and γ is the stress level below which the fracture probability is zero. In the case of brittle materials, there is no non-zero stress level for which it can be claimed that the material will not fail, so γ can be assumed to be zero.

Equation (5) can be rewritten as

$$\ln\left\{\ln\left[\frac{1}{1-F(x)}\right]\right\} = m\ln x - \ln \alpha \tag{8}$$

If X, Y and B are defined as follows, Eq. (8) is expressed as a linear equation, $Y = mX - B$, on log–log scales:

$$Y = \ln\left\{\ln\left[\frac{1}{1-F(x)}\right]\right\}$$

$$X = \ln x \tag{9}$$

$$-B = -\ln\alpha$$

$F(x)$ corresponds to the cumulative probability, P_i, which can be calculated from, for example, the symmetrical sample cumulative distribution (symmetrical ranks) expressed as

$$P_i = \frac{i-0.5}{N}, \quad i = 1, 2, \ldots, N \tag{10}$$

where N is the total number of strength data obtained under the same experimental conditions. When fitting the relation between

$$\ln\left(\ln\left\{\frac{1}{1-P_i[=F(x)]}\right\}\right)$$

and $\ln x$ from Eq. (8), we can derive the Weibull modulus and the scale parameter of the material.

Figure 4.12 shows the relationship between the fracture probability and the fracture strength of single-crystal Si specimens on a logarithmic scale. The nominal dimensions of the samples are listed in Table 4.1. The graph in Figure 4.12

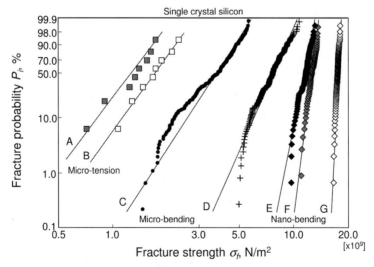

Figure 4.12 Weibull distribution of fracture strength for single-crystal silicon specimens at the micro/nano-level. Specimen size and deformation mode affect the strength. The slope of each line represents a scatter level of strength.

Table 4.1 Nominal dimensions of specimens used and
Weibull parameters measured from Figure 4.12.

Deformation mode	Specimen	Nominal dimension of specimen				Weibull parameter		Mean fracture strength × 10^9 (N m^{-2})
		Length (μm)	Upper width (μm)	Lower width (μm)	Thickness (μm)	Shape parameter	Scale parameter × 10^9 (N m^{-2})	
Tension	A	3000	250	250	24.5	3.88	1.41	1.32
	B	100	76	76	4	4.67	1.83	1.71
Bending	C	360	48	74.5	19	4.2	4.14	3.7
	D	36	4.85	7.5	1.9	7.24	8.09	7.68
	E		0.8	0.98		16.17	12.06	11.65
	F	6	0.55	0.72	0.25	23.04	13.08	12.72
	G		0.2	0.37		63.04	17.76	17.63

describes the fraction of Si specimens that fracture at different applied stresses, specimen sizes and deformation modes. Plots A and B illustrate the strength of micro-size specimens obtained from uniaxial tensile tests and other plots indicate that obtained with micro/nano-size specimens from doubly clamped bending tests [2, 21]. All the data can be fitted to respective straight lines. The slopes of these lines are closely related to the variability of strength (the Weibull modulus) for each specimen and tend towards the vertical with decreasing specimen size. The Weibull modulus of specimen A is found to be 3.88, which is approximately 6.1% of that of specimen E. This means that very little variation is observed in the strength of the smallest specimen. As compared with larger specimens, a smaller Si specimen displays greater strength, suggesting that a lesser number of small flaws, produced on the specimen surface through fabrication processes, is likely to lead to failure.

In addition to differences in specimen size and deformation mode, we must investigate the influence of specimen fabrication processes on the strength. The shape of a microstructure made of Si is usually produced by wet or dry etching that roughens an etched surface. A roughened surface has a considerable effect on the strength [22]. In wet etching, especially anisotropic etching of Si, solutions of potassium hydroxide (KOH), tetramethylammonium hydroxide (TMAH) and ethylenediamine pyrocatechol (EDP) are usually employed. These etchants form different etched surfaces, resulting in different specimen strengths. Dry etching, typically conducted by deep reactive ion etching (D-RIE), can fabricate a deep trench structure with a high aspect ratio, but makes an etched sidewall surface an irregular shape of scallops or ripples. The roughened surface plays a role as a notch where stress is concentrated that eventually leads to failure. The topology on the etched surface depends on the dry etching conditions. A surface treated by dry etching differs from that treated by wet etching. Therefore, Si structures that are fabricated using different etching processes yield variability of strength.

4.3.2
Metallic Materials

In the MEMS and semiconductor fields, many metallic materials are of use as structural components, such as electrical wiring, etching masks or sacrificial layers. For example, electroplated Ni is often utilized as a mold material in the "lithographie galvanoformung abformung" (LIGA) process that provides mass-produced high aspect ratio structures via embossing or injection micromolding of a wide variety of materials. Electroplated Ni is also used for the manufacture of microconnectors and microconductive probes in addition to the micromold. Metallic film materials exhibit a ductile deformation behavior at room temperature. Therefore, when using a ductile metal film as a structural material in miniature components, the plastic and creep properties coupled with elastic properties should be taken into account. The stress–strain trajectory during deformation should be formulated, allowing for optimum design of components.

A simple description of the plastic constitutive relationship of metals often employs the total strain increment theory, which avoids high-order term material constants. This assumes that the total strain increment, $d\varepsilon_t$, is simply expressed as the sum total of the elastic, plastic and creep strain increments, $d\varepsilon_e$, $d\varepsilon_p$ and $d\varepsilon_c$, respectively:

$$d\varepsilon_t = d\varepsilon_e + d\varepsilon_p + d\varepsilon_c \tag{11}$$

Considering a uniaxial stress state, the elastic strain component based on Hooke's law is given by

$$d\varepsilon_e = \frac{d\sigma}{E} \tag{12}$$

where $d\sigma$ is the stress increment and E is the Young's modulus. On the basis of the plastic potential theory, the plastic strain increment can be calculated as follows:

$$d\varepsilon_p = \frac{\partial f}{\partial \sigma} d\lambda \tag{13}$$

$$d\lambda = \frac{3d\bar{\sigma}}{2H'\bar{\sigma}} = \text{constant} \tag{14}$$

where f is the von Mises potential function, $\bar{\sigma}$ is the von Mises equivalent stress and H' is the work hardening modulus, defined as the slope of the relationship between the von Mises equivalent stress and the von Mises equivalent plastic strain, that is, $H' = d\bar{\sigma}/d\bar{\varepsilon}_p$. H' is also expressed as

$$H' = nA\varepsilon_p^{n-1} \tag{15}$$

where n is the work-hardening exponent, A is the plastic modulus and ε_p is the plastic strain. If the relationship between the stress increment, $\sigma - \sigma_y$, in the plastic region and ε_p is linear on the logarithmic scale, then the two-parameter Ludwik equation [23], expressed as follows, can be employed to estimate A and n:

$$\varepsilon_p = \left(\frac{\sigma - \sigma_y}{A} \right)^{1/n} \tag{16}$$

where σ_y is the yield stress. The tangent modulus, E_T, indicated by the slope of the stress–plastic strain relation after yield, is given by

$$E_T = \frac{EH'}{E + H'} \tag{17}$$

Figure 4.13 depicts the typical stress–strain relationships of electroplated Ni film from room temperature to 573 K [24]. All the specimens were tensioned at a constant strain rate of $1.0 \times 10^{-4}\,\mathrm{s^{-1}}$ until failure. The solid lines are plotted with reference to the axial strain and the open plots represent out-of-plane strain derived from XRD measurements in the elastic region. The Ni specimens exhibit elastic–plastic deformation until ductile failure for the entire range of temperatures examined. Tensile strength decreases with elevated temperature, whereas the break elongation increases with increasing temperature. In the elastic deformation

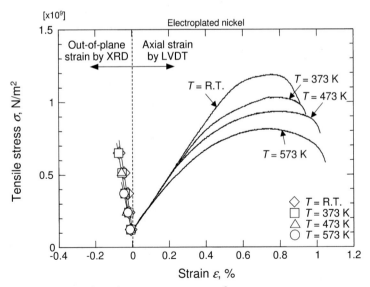

Figure 4.13 Typical tensile stress–strain curves of electroplated nickel film at intermediate temperatures. All the specimens exhibit ductility and the break elongation increases at elevated temperature.

range, the gradient between tensile stress and axial/out-of-plane strain gradually decreases with increasing temperature; therefore, the temperature during testing evidently influences the elastic–plastic deformation of the electroplated Ni specimen. The Young's modulus at room temperature averages 190.1 GPa, comparable to reported values, but is 8.2% smaller than that of bulk coarse-grained Ni. A constant Poisson's ratio of 0.23 is observed over the temperature range measured, but the value is approximately 20% smaller than that observed for the bulk material. Electroplated Ni films often include many vacancies inside the films that are produced during the electroplating process, so that the film density is lower in comparison with the bulk. A lower density would result in smaller elastic properties. The Young's modulus and the yield stress decrease with increasing temperature at a constant rate of −0.106 and −0.00184 GPa K^{-1}, respectively.

Figure 4.14 shows a typical relationship between the stress increment and plastic strain of electroplated Ni film. The yield stress was determined using a 0.2% offset method. The linear stress increment–plastic strain relation is observed over the temperature range examined. Consequently, A and n can be determined from the y-intercept at $\varepsilon_p = 10^0$ and the positive gradient of the linear relationship, respectively. By correlating the material constants, n, A, H' and E_T, with temperature, stress–strain curves at various temperatures can be generated as exemplified in Figure 4.15.

Most metals exhibit creep, which is a time-dependent permanent deformation resultant of constant stress at elevated temperatures. The creep test, used to determine the creep characteristics of a material, measures the resistance of a material

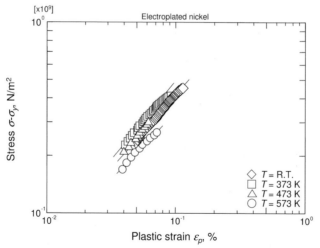

Figure 4.14 Relationship between the stress increment after the yield point and plastic strain in an electroplated nickel specimen. The relationship is shown as linear on the log–log scales, so that fitting using Ludwik's equation can be applied for determination of the material constants.

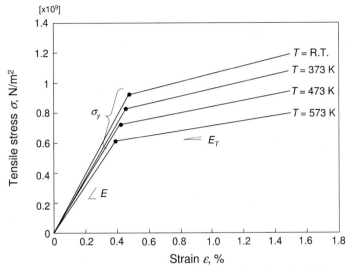

Figure 4.15 Estimated stress–strain relationship for electroplated nickel films.

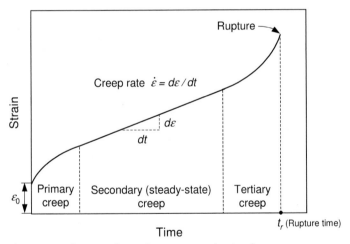

Figure 4.16 Schematic of typical creep curves showing the three creep stages: primary, secondary and tertiary creep. The creep rate during the secondary creep region is constant.

to deformation and failure when a specimen is subjected to a constant stress below the yield stress at an elevated temperature. The creep test has important significance from an engineering design perspective. A creep curve, which is the relationship of strain to time as illustrated in Figure 4.16, is divided into regions of primary (transient), secondary (steady-state) and tertiary creep. In the primary region, after the setting in of instantaneous elastic strain, ε_0, the material deforms rapidly, but at a decreasing rate. This deformation occurs with the contribution of dislocations "climbing" away from obstacles. In the secondary region, the rate at which dislocations move away from obstacles is almost equal to that at which dislocations are interrupted by other imperfections. The steady-state creep strain rate reaches a minimum value and remains approximately constant over a relatively long period of time. Eventually, in the tertiary region, specimen necking begins and at the same time the stress increases and the specimen deforms at an accelerated rate until rupture.

The generic mathematical expression of the creep curve is given as

$$\varepsilon_c = F(\sigma, T, t) = f(\sigma)g(T)h(t) \tag{18}$$

where σ, T and t are the stress, temperature and time, respectively. Time and temperature are explicitly included in the representation because these two parameters play key roles in the creep response. The above expression is based on the equation of state formulation utilized in the study of creep phenomena under variable stress. Often, as the simplest approach, only the steady-state creep is considered in the building of a creep constitutive relationship. The steady-state creep constitutive equation of Norton's rule, which is a good approximation for engineering metals, takes the following form:

$$\dot{\varepsilon}_{c2} \equiv \frac{d\varepsilon_{c2}}{dt} = A\sigma^n \tag{19}$$

where $\dot{\varepsilon}_{c2}$ is the steady-state creep strain rate, ε_{c2} is the steady-state creep strain and A and n are temperature-dependent material constants. These material constants are determined by fitting the steady-state creep region obtained under various constant stresses and temperatures. However, the creep strain rate actually decreases during creep and this is especially obvious at creep initiation in the primary region. This phenomenon is called creep hardening and is represented as a function of time or creep strain. In a design for structural integrity, the primary and secondary creep regions are not insignificant. A commonly used equation for the state representation of these two creep regions is provided by the Bailey–Norton law, which is given by

$$\varepsilon_{c12} = A\sigma^n t^m \tag{20}$$

where ε_{c12} is the primary and steady-state creep strain and m is a temperature-dependent material constant, as are A and n. Taking the derivative of ε_{c12} with respect to time produces the creep strain rate, $\dot{\varepsilon}_{c12}$, in the primary and secondary regions:

$$\dot{\varepsilon}_{c12} = A\sigma^n m t^{m-1} \tag{21}$$

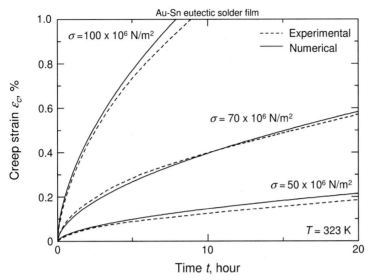

Figure 4.17 Creep strain curves for sputtered gold–tin film specimens at various constant stresses.

This is called the time hardening formulation of power law creep. Substitution of Eq. (20) into Eq. (21) provides the strain hardening formulation of power law creep:

$$\dot{\varepsilon}_{c12} = A^{1/m} m \sigma^{n/m} \varepsilon_c^{(m-1)/m} \tag{22}$$

The time and strain hardening formulations are employed to predict the creep curve under various stresses. Prediction of the creep curve is performed using the data obtained from multiple creep tests at each constant stress.

Figure 4.17 shows typical creep curves of an Au–Sn sputtered eutectic solder film at various constant stresses. The solid lines signify the estimated numerical creep curves and the dashed lines signify the representative experimental data. A good approximation to the tested creep curves is given by the estimated curves and the creep behavior at the primary and secondary creep stages is predictable using the time or strain hardening formulation.

4.3.3
Polymer Films

In general, polymers are considered to be more resistant to acidic and alkaline solutions than most metals. In the MEMS field, polymers are most always employed as a photoresist in the fabrication of MEMS devices and also as sealants in packaging. Recently, polymer films have played a role not only as these types of functional materials, but also as structural materials for MEMS. This is because a number of polymers make it possible to fabricate thick structures using relatively simple processes, such as high aspect ratio microstructure technology (HARMST). Such structures are found in various applications. For example, a microscale structure made of SU-8 polymer, which is usually used as a UV thick photoresist with negative transfer characteristics, serves as a flexible torsion beam in a two degrees of freedom mirror device [25]. The reason why SU-8 polymer has been employed is accounted for by the large tilt angle achieved by the micro-mirror with only a small external force. If a fluctuating or static load is applied to the SU-8 torsion beams for a long time, creep or stress relaxation of the SU-8 polymer will surely occur because of its viscoelasticity. Therefore, for the safe and durable operation of microcomponents made of polymers, material characteristics such as viscoelastic behavior should be evaluated.

Most thermoplastics exhibit non-Newtonian viscoelastic behavior, that is, the uniaxial stress and strain are not linearly related for most parts of the stress–strain curve. Such viscoelastic behavior means that both elastic and plastic (or viscous) deformation is generated when an external force is applied to a thermoplastic polymer. Figure 4.18 shows representative tensile stress–strain curves for the thermoplastic SU-8 polymer at various temperatures [26]. SU-8 polymer specimens of 1 mm length, 100 μm width and 10 μm thickness were heat treated at 468 K for 30 min. All the specimens were tensioned at a constant strain rate of $1.0 \times 10^{-4} \mathrm{s}^{-1}$. The SU-8 specimens exhibit elastic–inelastic behavior over the entire temperature range up to 473 K. The specimen tested at room temperature frac-

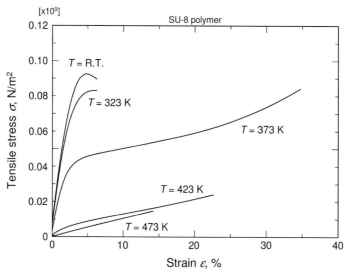

Figure 4.18 Typical tensile stress–strain curves for SU-8 polymer specimens. The deformation behavior is strongly influenced by temperature.

tured at a longitudinal strain of ~6%, whereas at 373 K specimen failure occurred at ~35%. The Young's moduli are found to average 3.6 GPa at room temperature and 0.1 GPa at 423 K. The tensile strength at room temperature is 0.95 GPa and shows a considerable decrease with increasing temperature. The stress–strain curves indicate that the uniaxial deformation of the SU-8 polymer shows a strong temperature dependence and the deformation behavior of SU-8 can be classified into two different types at the temperature boundary between 373 and 423 K. This is due to the glass transition temperature, T_g. Below T_g, many thermoplastic polymers are hard, brittle and glass-like. The glassy polymers have good strength, stiffness and creep resistance, but they have poor ductility and formability. On the other hand, above T_g, the polymers behave like a rubber. Therefore, if a thermoplastic polymer is utilized as a structural material in a MEMS device, the temperature must be maintained below T_g, allowing the polymer to deform in a glass-like manner.

To predict the overall shape of the stress–strain curves at various temperatures below T_g, the Ramberg–Osgood equation is often considered, as follows:

$$\varepsilon_t = \frac{\sigma}{E} + a\left(\frac{\sigma}{b}\right)^c \tag{23}$$

where E is the Young's modulus, σ is the stress, ε_t is the total strain, a is a constant and b and c are material constants representing strain rate and temperature dependence, respectively. Figure 4.19 depicts the stress–strain curves of SU-8

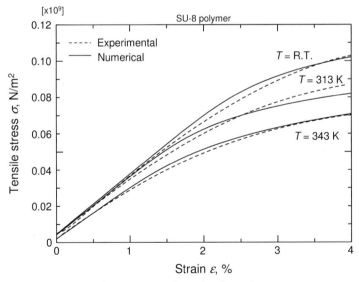

Figure 4.19 Estimated stress–strain relationship for SU-8 polymer specimens from the Ramberg–Osgood equation (solid lines) compared with those from tensile tests (dashed lines).

polymer at various strain rates at various temperatures. The solid lines indicate the predicted relationship using the Ramberg–Osgood equation and the dashed lines represent experimental data from uniaxial tensile tests. The calculated stress–strain curves at different strain rates and temperatures are in suitable agreement with the experimental data. This indicates that by assuming a uniaxial strain rate and temperature, the stress–strain curve of SU-8 polymer is predictable.

Thermoplastic polymers give rise to stress relaxation. Stress relaxation, like creep, is a reduction of the stress over a period of time at a constant strain due to viscoelastic deformation and can be measured in a steady environment by a stress relaxation test. The rate at which stress relaxation occurs is related to the relaxation time, which is considered to be a property of the polymer. The stress at time t after a stress is applied is expressed as

$$\sigma = \sigma_0 \exp\left(-\frac{t}{\lambda}\right) \tag{24}$$

where σ_0 is the initial stress and λ is the relaxation time. The relaxation time is dependent on the viscosity and temperature:

$$\lambda = \lambda_0 \exp\left(\frac{Q}{RT}\right) \tag{25}$$

where λ_0 is constant, R is the gas constant and Q is the activation energy. The activation energy represents the ease with which the polymer chains slide past each other.

To investigate viscoelastic behavior of polymers, the relaxation modulus can be measured. The following equation represents the relaxation modulus, E_r, for polymers:

$$E_r(t, T) = \frac{\sigma(t, T)}{\varepsilon_0} \tag{26}$$

where t is the absolute time of the stress relaxation test and ε_0 is the initial strain. The time–temperature equivalence assumes that the viscoelastic behavior of a polymer at one temperature can be related to that at another temperature by only a change in the time scale, as depicted schematically in Figure 4.20. Therefore, the relaxation modulus–duration time curves provide a single master curve that can be shifted along the time axis. The shifted relaxation moduli are no longer dependent on absolute time and temperature, but are instead a function only of reduced time, t_R. The reduced time for construction of a master curve simply defines the time–temperature shift factor, $a_{T_0}(T)$:

$$a_{T_0}(T) = \frac{t}{t_R} \tag{27}$$

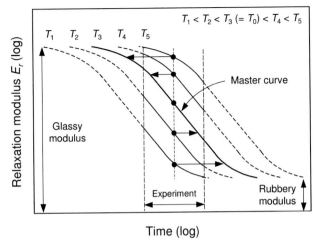

Figure 4.20 Schematic of the master curve derivation for the relaxation modulus. Under the assumption of time–temperature equivalence, the viscoelastic behavior of polymers at one temperature can be related to that at another temperature by only a change in the time scale.

The relaxation modulus in a master curve that can be represented by a Prony series, based on the generalized Maxwell model:

$$E_r(t) = E_\infty + \sum_{i=1}^{N} E_i e^{-t/\tau_i} \tag{28}$$

where E_∞ is the rubbery asymptotic modulus, E_i are the Prony series coefficients and τ_i are the relaxation times.

Figure 4.21 represents the master curve for the relaxation modulus of SU-8 polymer along with its stress relaxation curves. The data were fitted with a Prony series by means of a collocation method. The measured relaxation modulus is plotted against the converted logarithmic time scale at the standard temperature, T_0, of 413 K. The relaxation moduli of the SU-8 polymer produce a smooth master curve over the entire range of converted times and the approximation curves from the Prony series also have smooth curve shapes. The master curve has an inflection region between 373 and 433 K, in which the glass–rubber transition occurs. The estimated values of the glassy–rubbery moduli obtained from the curve are 3.2 and 0.12 GPa, respectively; almost the same as those from the stress–strain curves. The construction of a master curve allows the optimum design of MEMS components by finite element analysis with consideration of time-dependent stress relaxation.

Assuming that the shift factor has a functional form of Arrhenius type, the apparent activation energy, ΔH, is given by

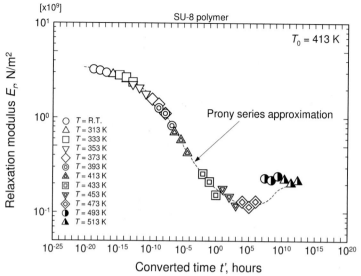

Figure 4.21 Master curve of the relaxation modulus for an SU-8 polymer specimen. The master curve exhibits a smooth "S" shape.

$$\log a_{T_0} = \frac{\Delta H}{2.303R}\left(\frac{1}{T} - \frac{1}{T_0}\right)$$

(29)

When the shift factor is plotted on the common logarithmic scale against the inverse absolute temperature, the slope of the straight line provides the apparent activation energy.

Figure 4.22 shows the relationship between the shift factor and inverse temperature. The shift factor of a microscale SU-8 polymer specimen is close to that of milli-scale specimens over the entire temperature range. The figure includes two approximately straight lines bisected at an approximate T_g of 413 K. The apparent activation energy of two different-sized SU-8 polymer specimens below T_g is 258 kJ mol^{-1}, which is comparable to those for typical epoxy resins. Both specimens have a similar value of the apparent activation energy for each state. Consequently, the influence of specimen size on flowability, that is, the ease of segmental motion in the SU-8 polymer, is small.

4.3.4
Shape Memory Alloy Films

For one variety of integrated MEMS actuator mechanisms, shape memory alloy (SMA) films are promising candidates due to their higher work output per unit volume compared with other actuation mechanisms, such as bimetallic, electro-

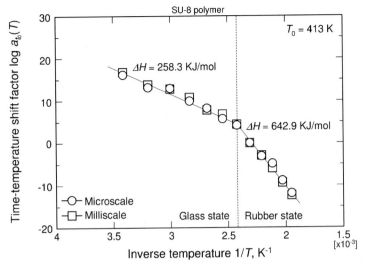

Figure 4.22 Variation of the time–temperature shift factor with inverse temperature. The slope indicates the apparent activation energy.

static and piezoelectric actuators. Several applications of SMA MEMS actuators have recently been presented, such as a micropump, microgripper and a probe card [27]. Representative examples of SMA films are Ti–Ni, Cu–Al–Ni and Fe–Pd [28]. In particular, Ti–Ni film is the most popular SMA film in the MEMS field, due to the ease of preparation by sputtering and fabrication with etching techniques. The shape memory and mechanical performance of these SMA films is well known to exhibit film composition and heat treatment process dependence. Precise understanding of the thermomechanical behavior is critical for the full exploitation of a material's potential and is expected to lead to the design and development of high-performance SMA film-actuated MEMS.

First, in the design of SMA film-actuated MEMS devices, the stress–strain relationship of the SMA film must be investigated. Assuming that only a uniaxial stress state is considered and no plastic deformation occurs, the thermomechanical constitutive equation for polycrystalline SMA materials is expressed as

$$\dot{\sigma} = E\dot{\varepsilon} + \Theta\dot{T} + \Omega\dot{\xi} \tag{30}$$

where σ, ε, T, E, Θ and Ω are the stress, strain, temperature, modulus of elasticity, thermoelastic modulus and the transformation expansion constant, respectively. The dot over a symbol denotes the material derivative. ξ is the volume fraction of the martensitic phase, that is, $\xi = 0$ indicates an austenitic phase and $\xi = 1$ indicates a martensitic phase.

In the case of isothermal deformation processes ($\dot{T} = 0$) during transformation, the stress increment ranging from the onset time ($t = 0$) until the end

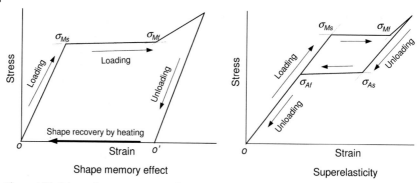

Figure 4.23 Schematic representation of stress–strain curves showing the shape memory effect and superelasticity for shape memory alloy materials.

$(t = t^*)$ of the martensitic transformation is expressed by modifying Eq. (30) as follows:

$$d\sigma = \int_0^{t^*} E\dot{\varepsilon}dt + \int_0^{t^*} \Omega\dot{\xi}dt \tag{31}$$

Assuming that E and Ω are constants, the loading process during transformation from austenite to martensite, as shown schematically in Figure 4.23, can be expressed by rewriting Eq. (31) as

$$d\sigma = Ed\varepsilon + \Omega, \quad d\sigma > 0, \quad d\varepsilon > 0 \tag{32}$$

where $d\varepsilon$ [$= \varepsilon(t^*) - \varepsilon(0)$] is the strain increment corresponding to $d\sigma$. The unloading process from martensite to austenite can also be expressed as follows, because the volume fraction of martensite changes from 1 to 0:

$$d\sigma = Ed\varepsilon - \Omega, \quad d\sigma < 0, \quad d\varepsilon < 0 \tag{33}$$

The loading or unloading process in the martensite or austenite phase can be given by

$$d\sigma = Ed\varepsilon \tag{34}$$

which is obtained because the volume fraction is unchanged due to no transformation ($\dot{\xi} = 0$).

Figure 4.24 shows typical tensile stress–strain curves for Ti–Ni SMA films of various composition at room temperature and 353 K [29]. Ti–Ni films were deposited by sputtering of a controlled composition. At room temperature, the Ni–48.0 at.% Ti film exhibits superelasticity; that is, residual strain is not observed

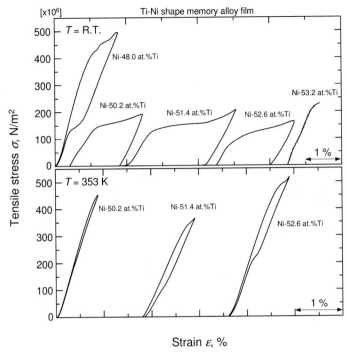

Figure 4.24 Typical tensile stress–strain curves for titanium–nickel shape memory alloy films. The deformation behavior is wholly affected by the film composition and temperature.

after unloading due to the presence of the austenite phase. In films with a Ti content ranging from 50.2 to 52.6 at.%, the tensile stress–strain curves exhibit shape memory behavior due to the martensite phase. Residual strain is eliminated by heating after unloading. The Ni–53.2 at.% Ti film has fractured at a longitudinal strain of only 0.8%. The brittle Ti-rich film has fractured at a small strain, because of precipitation hardening brought on by an increase in the Ti content. At 353 K, Ti–Ni films with Ti contents of 50.2–52.6 at.% exhibit superelastic behavior. These observations indicate that both film composition and ambient temperature affect the thermomechanical behavior of sputtered Ti–Ni films. The Young's modulus and the transformation stresses, σ_{M_s}, σ_{M_f}, σ_{A_s} and σ_{A_f}, can be determined from the slope and the inflection points of the stress–strain curve, respectively.

In transformation phenomenology, the transformation kinetics for the martensitic transformation are determined using the Koistinen–Marburger and Magee equations:

$$\frac{\dot{\xi}}{1-\xi} = b_M c_M \dot{T} - b_M \dot{\sigma}, \quad b_M c_M \dot{T} - b_M \dot{\sigma} \geq 0 \tag{35}$$

and for the reverse transformation

$$\frac{\dot{\xi}}{\xi} = b_A c_A \dot{T} - b_A \dot{\sigma}, \; b_A c_A \dot{T} - b_A \dot{\sigma} \geq 0 \tag{36}$$

where b_M, c_M, b_A and b_M are the material constants, which generally depend on the stress and temperature. Integration of Eqs. (35) and (36), from the initial values ($\sigma = 0$, $T = M_s$ or A_s) to the current values (σ, T, ξ), leads to the explicit kinetic forms for the martensitic and reverse transformations, respectively:

$$\xi = 1 - \exp[b_M c_M (M_s - T) + b_M \sigma], \; b_M c_M (M_s - T) + b_M \sigma \leq 0 \tag{37}$$

$$\xi = \exp[b_A c_A (A_s - T) + b_A \sigma], \; b_A c_A (A_s - T) + b_A \sigma \leq 0 \tag{38}$$

where M_s and A_s are designated as the transition temperatures for the onset of the martensitic and reverse transformations at zero stress, respectively. Based on the assumption that the martensitic transformation is completed at $\xi = 0.99$, the stresses that induce the martensitic transformation are defined by rewriting Eqs. (37) and (38) as follows:

$$\sigma_{M_s} = C_M (T - M_s) \tag{39}$$

$$\sigma_{M_f} = -\frac{2 \ln 10}{b_M} + c_M (T - M_s) \tag{40}$$

where σ_{M_s} and σ_{M_f} are the critical stresses at the onset and at the end of the martensitic transformation, respectively. Replacing the subscript M with A gives the expression for the reverse transformation. Figure 4.25 shows the variation in the critical stress of an Ni–52.6 at.% Ti film with respect to temperature. The transformation stress is given as a linear function of temperature. c_M and c_A are calculated from the slope of the transformation stress-temperature relation. The increasing rates of the stresses, σ_{A_s} and σ_{A_f}, with increase in temperature are higher than those of σ_{M_s} and σ_{M_f}, indicating a greater influence of temperature on the reverse transformation. At constant temperature, the difference in critical stress between the onset and the end of the transformation provides the equation for martensitic transformation:

$$d\sigma_M = -\frac{2 \ln 10}{b_M} \tag{41}$$

and the equation for the reverse transformation:

$$d\sigma_A = \frac{2 \ln 10}{b_A} \tag{42}$$

These equations imply that $b_M < 0$ and $b_A > 0$ and that b_M and b_A are the material constants related to the width of transformation. The material constants, b_M, b_A, c_M and c_A, obtained from uniaxial tensile tests, are shown as a function of Ti

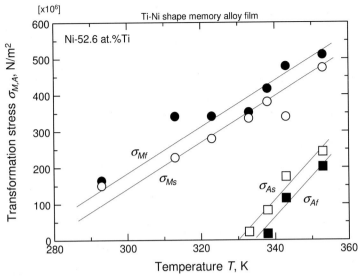

Figure 4.25 Transformation stress as a function of temperature showing a linear relationship. The relationship results in the derivation of material constants b_M, b_A, c_M and c_A.

content in Figure 4.26. All the constants are affected by the film composition. It is known that Ti-rich Ti–Ni films easily precipitate Ti_2Ni, which strongly affects the rearrangement processes during transformation. In addition, the constants measured for Ti–Ni films differ from the Ti–Ni bulk values. This is caused by the difference in the surface area-to-volume ratio. These constants are given as functions of the Ti content, C_{Ti}:

$$b_{M,A}, c_{M,A} = \gamma C_{Ti} + \eta \tag{43}$$

The fitted coefficients, γ and η, for Ti–Ni films are listed in Table 4.2.

Figure 4.27 compares the estimated stress–strain relations of a, Ni–52.6 at.% Ti film, as determined from the constitutive equations, with those measured from tensile tests. The solid and dashed lines indicate the estimated and experimental relationships, respectively. The estimated stress–strain curves show very close agreement with the experimental curves over the entire temperature range. This equivalence demonstrates the effectiveness of the constructed constitutive equations for use in the design of Ti–Ni film-actuated MEMS devices.

4.4

Fatigue Testing of Film Specimens

Fatigue is defined as the lowering of strength of a material due to a repetitive stress, which is often below the yield stress. MEMS devices, such as accelerome-

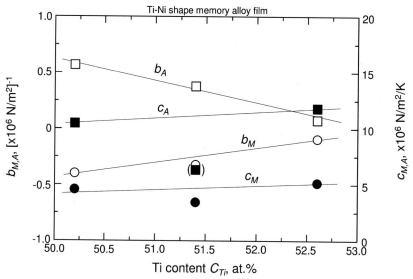

Figure 4.26 The material constants, b_M, b_A, c_M and c_A, obtained from uniaxial tensile tests as a function of Ti content. Film composition affects these constants.

Table 4.2 The fitted coefficients, γ and η, used in Eq. (43), for Ti–Ni films.

	b_M	b_A	c_M	c_A
γ	0.13	−0.21	0.35	0.56
η	6.86	10.9	−13.8	−19.5

ters, pressure sensors and optical scanners, often employ elastically deformable microscale hinge structures for the generation of cyclic mechanical motion. During operation, the microcomponents are routinely subjected to fluctuating mechanical stresses in the form of tension, compression, bending, vibration, thermal expansion and contraction. The stress yields an accumulation of fatigue damage, which is accompanied by the formation of inclined critical planes (such as the {111} plane in the case of single-crystal Si) as a means of dissipating energy during the progress of fatigue damage. When the cyclic motions reach a sufficient number of times, the microcomponents exhibit premature failure due to fatigue. Hence the investigations of fatigue damage behavior, and also the quasi-static mechanical behavior of microscale structures, are critical to the safe and reliable design of high-performance MEMS devices.

The fatigue test measures the number of loading cycles that need to be applied by a specific load until failure. Although a conventional and older method that has

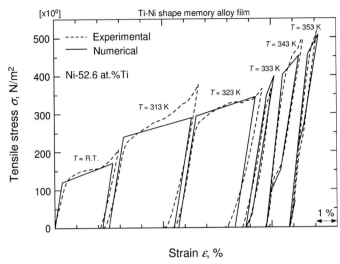

Figure 4.27 Comparison between the estimated stress–strain relations of Ni–52.6 at.% Ti film from the constitutive equations (solid lines) and those from tensile tests (dashed lines).

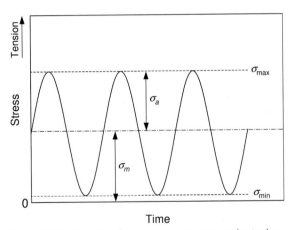

Figure 4.28 Schematic of tension–tension stress cycles in the uniaxial tensile mode fatigue test of a film specimen.

been employed for the measurement of macro-material resistance to fatigue is the rotating cantilever beam test, it is not suitable for specimens of micro- and nano-level size. A fatigue test that imparts cyclic stresses in the form of tension or bending is suitable for such specimens.

Figure 4.28 illustrates the stress cycles during a tensile-mode fatigue test. In this case, the cyclic loading varies between the defined maximum and minimum tensile stresses. Usually, when a test of a micro/nanoscale specimen is carried out, either a combination of stress amplitude plus mean stress or a combination of maximum stress plus minimum stress is adopted for stress control. The cyclic stresses are applied to the specimen at a constant frequency with a sine or triangle waveform. The stress amplitude, σ_a, and the mean stress, σ_m, are represented as follows:

$$\sigma_a = \frac{\sigma_{max} - \sigma_{min}}{2} \tag{44}$$

$$\sigma_m = \frac{\sigma_{max} + \sigma_{min}}{2} \tag{45}$$

where σ_{max} and σ_{min} are the maximum and minimum stresses during the test, respectively. As the mean stress increases, the stress amplitude must decrease for the material to bear the applied stresses. The Goodman relation expresses this condition as

$$\sigma_a = \sigma_{fs} \left[1 - \left(\frac{\sigma_m}{\sigma_T} \right) \right] \tag{46}$$

where σ_{fs} is the desired fatigue strength at zero mean stress and σ_T is the tensile strength obtained from a quasi-static tensile test. One of the most important features requiring measurement is a material's base stress, which is used for the determination of the maximum and minimum stresses during the fatigue test. The mean fracture strength of a material, obtained from the quasi-static test, is most often employed as the base stress. Essentially, the mean fracture strength must be measured under comparable experimental conditions before the fatigue test is conducted. The statistical calculations described in the previous section have important implications in the base stress determination. In addition, temperature and humidity affect the initiation and propagation of dislocations or cracks produced during the fatigue test. These environmental factors are of great significance and dominate the lifetime of a material. Therefore, the fatigue test should be performed within an environmental control unit.

Figure 4.29 exhibits a typical applied stress–time waveform obtained from a tensile-mode fatigue test of a single-crystal Si specimen [30]. The stress–time waveform of the specimen is configured with stable triangular waves with a small frequency deviation. The maximum stress and stress amplitude were kept constant throughout the test. An abrupt stress drop at 705 s is observed. This indicates that time-delayed fatigue failure has occurred at 35 250 cycles, which is equivalent to a lifetime of the specimen under these test conditions.

Figure 4.30 shows the relationship between the applied maximum stress and the number of cycles to failure, N_f, for single-crystal Si specimens [31]. The fatigue life of each specimen increases monotonically with decreasing peak stress. In other words, a half reduction in peak stress during the cyclic test would provide

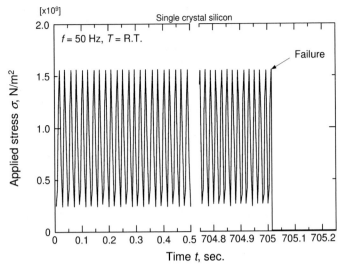

Figure 4.29 Example of stress cycles for a microscale single-crystal silicon specimen during cyclic loading.

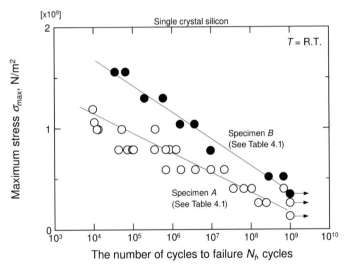

Figure 4.30 Stress–cycle number curves for microscale single-crystal silicon specimens. The detail of the specimen dimensions is listed in Table 4.1. Fatigue life is extended with decrease in the maximum stress, but no fatigue limit is observed.

an increase in the life by as much as two orders of magnitude. The Si specimen does not have a manifest fatigue limit, in that the *S–N* curves continue their downward trends at greater N_f values, although a few specimens have not failed even after gigacycles. Additionally, the effect of specimen size on the fatigue life is clearly observed between the different sized specimens. This effect is presumably due to the smaller number of latent defects affecting fatigue failure at the surface, in the case of a small specimen. At the same stress state, a smaller size Si specimen has better durability to fluctuating stresses than a larger specimen.

4.5
Conclusion

In this chapter, we have presented technical information regarding the uniaxial tensile testing of micro- and nanoscale specimens. The major technical difficulties in conducting a tensile test at the micro/nanoscale are considered to be specimen chucking and deformation measurement. In the near future, there will possibly be no more innovative techniques addressing these two concerns. As a prospective study, upgrading of the chucking method in terms of precision and ease is important. The on-chip uniaxial tensile test is an excellent method, because the special techniques required for alignment, chucking and handling are not necessary for a micro- or nanoscale specimen. The precise fabrication of on-chip devices is also one of the strong points of MEMS technology.

A method was introduced for obtaining a constitutive relationship for MEMS materials by means of the uniaxial tensile test. Many types of materials, such as metals, polymers, hard coat films, functional films and ceramics, are applicable to constituting MEMS. Their characteristics show considerable differences between each other. To improve the safety and durability of MEMS devices, the mechanical characteristics of the micro/nanoscale materials must first be quantified in a practical environment, so that the characteristics inherent in these materials can be exploited in the structural design of MEMS devices. The uniaxial tensile test is a standard method for the evaluation of material properties in the MEMS field and therefore better design, based on truly representative material properties measured by the tensile test, can lead to the development of safe and durable MEMS devices that exhibit better performance.

References

1 T. Yi, C. J. Kim, *Meas. Sci. Technol.* **10** (1999) 706–716.

2 T. Namazu, Y. Isono, T. Tanaka, *J. Microelectromech. Syst.* **9** (2000) 450–459.

3 H. Ljungcrantz, L. Hultman, J.-E. Sundgren, S. Johansson, N. Kristensen, J.-Å. Schweitz, *J. Vac. Sci. Technol. A* **11** (1993) 543–553.

4 W. C. Oliver, G. M. Pharr, *J. Mater. Res.* **7** (1992) 1564–1583.

5 D. R. Askeland, P. P. Phule, *The Science and Engineering of Materials*, Brooks/Cole-Thomson Learning, Belmont, CA 2003.

6 T. Tsuchiya, O. Tabata, J. Sakata, Y. Taga, *J. Microelectromech. Syst.* **7** (1998) 106–113.

7 Y. Isono, T. Namazu, N. Terayama, *J. Microelectromech. Syst.* **15** (2006) 169–180.

8 T. Namazu, S. Inoue, H. Takemoto, K. Koterazawa, *IEEE Trans. Sens. Micromach.* **125** (2005) 374–379.

9 K. Sato, M. Shikida, Proc. IEEE Int. Conf. Microelectromech. Syst. (1996) 360–364.

10 H. Ogawa, K. Suzuki, S. Kaneko, Y. Nakano, Y. Ishikawa, T. Kitahara, Proc. IEEE Int. Conf. Microelectromech. Syst. (1997) 430–433.

11 W. N. Sharpe, Jr., B. Yuan, R. L. Edwards, *J. Microelectromech. Syst.* **6** (1997) 193–199.

12 I. Chasiotis, W. G. Knauss, *Exp. Mech.* **35** (2003) 217–231.

13 H. Huang, F. Spaepen, *Acta Mater.* **48** (2000) 3261–3269.

14 T. Ishida, K. Kakushima, M. Mita, H. Fujita, Proc. IEEE Int. Conf. Microelectromech. Syst. (2005) 879–882.

15 T. Namazu, S. Inoue, D. Ano, K. Koterazawa, Proc. IEEE Int. Conf. Microelectromech. Syst. (2004) 157–160.

16 T. Namazu, S. Inoue, D. Ano, K. Koterazawa, *Key Eng. Mater.* **297–300** (2005) 574–580.

17 M. Kiuchi, S. Matsui, Y. Isono, J. *Microelectromech.* Syst. **16** (2007) 191–201.

18 S. M. Spearing, *Acta Mater.* **48** (2000) 179–196.

19 K. E. Petersen, *Proc. IEEE* **70** (1982) 420–457.

20 H. J. McSkimin, *J. Appl. Phys.* **24** (1953) 988–997.

21 T. Namazu, Y. Isono, T. Tanaka, *J. Microelectromech. Syst.* **11** (2002) 125–135.

22 T. Yi and C.-J. Kim, Proc. Int. Conf. Solid-State Sens. Actuators (1999) 518–521.

23 P. Ludwik, *Elemente der Technologischen Meckanik*, Springer, Berlin 1909.

24 T. Namazu, S. Inoue, *J. Fatigue Fract. Eng. Mater. Struct.* **30** (2007) 13–20.

25 T. Fujita, K. Maenaka, Y. Takayama, *Sens. Actuators A* **121** (2005) 16–21.

26 T. Namazu, S. Inoue, K. Takio, T. Fujita, K. Maenaka, K. Koterazawa, Proc. IEEE Int. Conf. Microelectromech. Syst. (2005) 447–450.

27 T. Namazu, S. Inoue, Y. Tashiro, Y. Okamura, K. Koterazawa, Proc. Int. Conf. Solid-State Sens. Actuators Microsyst. (2005) 847–850.

28 S. Inoue, T. Namazu, S. Fujita, K. Koterazawa, K. Inoue, *Mater. Sci. Forum* **426–432** (2003) 2213–2218.

29 T. Namazu, S. Inoue, A. Hashizume, K. Koterazawa, *Vacuum* **80** (2006) 726–731.

30 T. Namazu, Y. Isono, Proc. IEEE Int. Conf. Microelectromech. Syst. (2003) 662–66.

5
On-chip Testing of MEMS

Harold Kahn, Department of Materials Science and Engineering, Case Western Reserve University, Cleveland, OH, USA

Abstrast

On-chip testing of miroelectromechanical systems (MEMS) involves fabricating the force actuators and the test specimens simultaneously. In this way, the attachments and alignments between the specimens and load cells will be perfect. Loading is typically done using applied voltages for electrostatic or thermal expansion actuation. However, as-fabricated residual stressed can also be used to generate forces. In this chapter, on-chip testing techniques are reviewed for measuring: Young's modulus, residual stresses, friction, wear, fracture toughness, fracture strength, and fatigue.

Keywords

on-chip testing; mechanical testing; MEMS

Reliability of MEMS.
Edited by O. Tabata, T. Tsuchiya
Copyright © 2013 WILEY-VCH Verlag GmbH & Co. KGaA, Weinheim
ISBN: 978-3-527-33501-5

5.1
Introduction to On-chip Mechanical Testing of MEMS

Mechanical testing of micromachined devices and "MEMS" can be performed using external loading devices. However, this involves significant difficulties in attaching and aligning the microtest specimens to the external loading system. The attachments must be extremely secure, because any slippages, even on the nanometer scale, would be catastrophic to the measurement. Similarly, a variation in alignment of a fraction of a micron can inadvertently change a tensile test into a bending test, rendering the test results invalid.

On-chip testing eliminates these potential problems. The test specimens and the loading devices are fabricated simultaneously on the same substrate, usually from the same materials. Therefore, the as-fabricated device contains the loading device and test specimen in perfect alignment. The specimen and loading actuator can also be attached as strongly as required, since the user-defined device design determines the connection between the two.

For on-chip testing, the only allowable inputs to the die are electrical connections. Therefore, any forces necessary to make the mechanical measurements must be generated using on-chip actuators. Typically, electrostatic actuators are employed, most often comb-drive actuators. One possible disadvantage of on-chip testing is the limit of the forces and displacements that can be generated using microfabricated actuators. In order to avoid arcing between the electrodes, applied voltages must be kept below about 200 V, when testing in air. However, as will be discussed in this chapter, sufficient forces and displacements can be achieved for virtually any mechanical test with an appropriate device design. For example, thermal expansion actuators have been used, which can generate large forces. Scratch drives can also create large forces and essentially unlimited displacements.

5.2
Young's Modulus Measurements

Before any other mechanical properties can be measured, the elastic response – Young's modulus – of the material must be established. The material's Young's modulus, E, relates stress, σ, and strain, ε, by the following equation:

$$\sigma = E\varepsilon \tag{1}$$

It will therefore dictate the material's force–displacement behavior. Typically, for on-chip testing, the displacement of the device is monitored during the experiment, either optically (through an optical or electron microscope) or electrically (by use of capacitive sensors, for example). If the Young's modulus of the material is known, the measured displacements can be converted into forces, which can in turn be used to determine stresses present at various parts of the devices. Finite element analysis (FEA) is often used to make these conversions and Young's modulus is a necessary input.

For bulk materials, Young's modulus is determined by measuring the force–displacement, or stress–strain, behavior of a specimen of known geometry in pure tension. This has frequently been done with MEMS materials using external load cells, since very large forces are necessary to create measurable displacements. Only recently [1] has a strong enough actuator, using parallel-plate electrostatic actuation, been designed to perform similar on-chip measurements. Typically, MEMS testing has been performed through less direct techniques.

5.2.1
Lateral Resonant Structures

The earliest on-chip test devices developed for Young's modulus determination are laterally driven resonant structures [2]. Two such devices are sketched in Figure 5.1, (a) with straight beams and (b) with folded beams. In both schematics, the anchors are represented by solid black features, and the cross-hatched features are fully released and free to move. These devices, like the majority of the on-chip

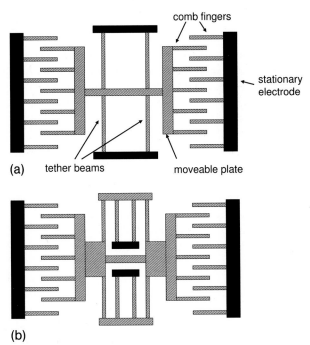

Figure 5.1 Schematic top views of lateral resonators with (a) straight tether beams and (b) folded tether beams. The solid black features are anchored to the substrate and the cross-hatched features are fully released.

devices discussed in this chapter, are most often fabricated from polycrystalline silicon (polysilicon). The folded-beam design was developed to accommodate residual stresses in the polysilicon structural film. For example, if the polysilicon contained substantial compressive residual stresses, the straight tether beams shown in Figure 5.1a would buckle and the moveable plate would not be able to move laterally (at least not smoothly). However, the folded tether beams in Figure 5.1b would simply expand slightly (away from the moveable plate) and would thus relieve their compressive residual stresses.

For both designs, the lateral resonator takes advantage of the relationship between the resonant frequency, f_r, and mass, M, of the moveable portion of the device and the stiffness of the tether beams:

$$f_r = \frac{1}{2\pi} \left(\frac{k_{sys}}{M} \right)^{1/2}$$ (2)

where k_{sys} is the spring constant of the tether beams. The spring constant is given by

$$k_{sys} = 24 EI / L^3$$ (3)

$$I = hw^3 / 12$$ (4)

where E is Young's modulus of the structural material (polysilicon), I is the moment of inertia of the tether beams and L, h and w are the length, thickness and width, respectively, of the tether beams. Therefore, by combining these equations and measuring f_r, E can be determined. (It is assumed that the density, and thus the mass, of the structural material are known.)

The resonant frequency is determined by driving the device with an a.c. voltage at various frequencies and monitoring the motion of the moveable plate. As shown in Figure 5.1, the device utilizes comb-drive actuators to generate the force on the moveable plate. When a voltage is applied across either set of interdigitated comb fingers shown in Figure 5.1a and b, an electrostatic attraction will be generated due to the increase in capacitance as the overlap between the comb fingers increases. The force, F, generated by the comb drive is given by

$$F = \frac{1}{2} \frac{\partial C}{\partial x} V^2 = n\varepsilon \frac{h}{g} V^2$$ (5)

where C is capacitance, x is distance traveled by one comb-drive toward the other, n is the number of pairs of comb fingers in one drive, ε is the permittivity of the fluid between the fingers (air), h is the height of the fingers, g is the gap spacing between the fingers and V is the applied voltage [2]. Typically, a relatively large d.c. bias is applied with a relatively small a.c. signal. This keeps the frequency of the force applied to the device equal to the frequency of the drive voltage, even though force is related to the square of the voltage. The frequency of the a.c. signal is swept over an appropriate range and when the a.c. frequency matches the resonance frequency, the moveable plate will vibrate. Thus, from this measurement and the above equations, Young's modulus can be determined.

If Young's modulus of one material is known, the lateral resonators can be fabricated from a two-layer composite and the resonance measurements can be used to determine Young's modulus of the second material [3, 4], assuming its density is known. Similarly, if the second material can be deposited on the moveable plate, but not on the tether beams, the change in resonant frequency will determine the mass (and density if the dimensions can be measured) of the second material [5]. For any measurement, it is highly desirable that measurements be taken from devices with varying tether beam lengths and/or tether beam widths, in order to minimize experimental inaccuracies.

5.2.2
Cantilever Measurements

Another on-chip technique for determining Young's modulus involves pulling downwards on a cantilever beam by means of an electrostatic force [6]. An electrode is fabricated into the substrate beneath the cantilever beam and a voltage is applied between the beam and the bottom electrode. The force acting on the beam is given by

$$F(x) = \frac{\varepsilon_0}{2} \left[\frac{V}{g + z(x)} \right]^2 \left\{ 1 + \frac{0.65[g + z(x)]}{w} \right\}$$

(6)

where $F(x)$ is the electrostatic force at x, ε_0 is the dielectric constant of air, g is the gap between the beam and the bottom electrode, $z(x)$ is the out-of-plane deflection of the beam, w is the beam width and V is the applied voltage; this equation has been corrected to include the effects of fringing fields acting on the sides of the beam [6]. The deflection of the cantilever beam as a function of position can be accurately measured using optical interferometry and then used to determine the force–displacement (stress–strain) behavior for the cantilever beam. Young's modulus can then be found.

An extension of this technique uses fixed–fixed beams instead of cantilever beams [7] (a fixed–fixed beam is one that is anchored on both ends). A schematic of this geometry with an example of the interferometry measurement is shown in Figure 5.2. Again, using an applied voltage between the beam and an underlying electrode, the force–displacement behavior is determined.

5.3
Residual Stress Measurements

When the device contains fixed–fixed beams, as shown in Figures 5.1a and 5.2, the force–displacement behavior of the device will depend on the residual stress in the beams, in addition to Young's modulus. In this case, using both fixed–fixed

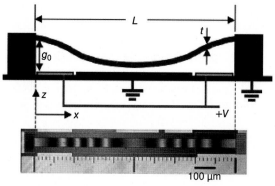

Figure 5.2 Schematic side and top view interferogram showing a fixed–fixed beam deflected downward at its center by the application of a voltage between the beam and an electrode on the substrate [7].

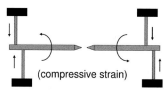

Figure 5.3 Schematic top view of a rotating strain gauge. The solid features are anchors and the arrows indicate the direction of expansion and rotation upon release for a material containing a compressive residual stress.

beams and also folded beams or cantilevers will provide both Young's modulus and residual stress. However, there are simpler devices that are specifically designed to measure residual stresses.

5.3.1
Passive Stress Measurement Devices

Several devices have been developed to determine compressive residual stresses based on the buckling of beams [8, 9], but the simplest device design for measuring residual stresses is a rotating strain gauge, as sketched in Figure 5.3 [10]. In this schematic, the black pads indicate the anchors. Since the two anchored legs of the device are offset where they attach to the central pointing beam, when released, the central beam will rotate as the legs expand or contract to relieve any residual compressive or tensile stresses. In the figure, the effects of a compressive residual stress are indicated by the arrows. Since there are two opposing strain gauges in Figure 5.3, the displacement at the ends of the pointing beams will be doubled relative to each other and the sensitivity of the device is therefore increased. Measuring the deflection of the device after release, in conjunction with FEA, will reveal the magnitude and sign of the residual stress.

5.3.2
Stress Gradients

For structures fabricated from thin deposited films, the stress gradient can be equally as important as the stress itself. The easiest structure for measuring stress gradients is a simple cantilever beam. By measuring the end deflection, d, of a cantilever beam of length l and thickness t, the stress gradient, $d\sigma/dt$ is determined by [11]

$$\frac{d\sigma}{dt} = \frac{2\delta}{l^2} \frac{E}{1-\nu} \tag{7}$$

where ν is Poisson's ratio. The magnitude of the end deflection can be measured by microscopy, optical interferometry or any other technique.

5.4
Stiction and Friction

Stiction of MEMS is a common reliability concern. The appearance of stiction and the improvements attained by various treatments are typically quantified by simple cantilever beams of increasing length [12]. As the length of a cantilever beam increases, its flexibility perpendicular to the substrate increases and its susceptibility to stiction is enhanced. Optical interferometry can readily detect stiction of a cantilever beam and determining the critical beam length at which stiction occurs will reveal the susceptibility to stiction.

Friction is another mechanical property that is related to the forces between contacting parts. An early on-chip friction measurement [13] used a comb-drive actuator to displace a moveable plate. Then, a voltage applied between the plate and an electrode on the substrate created a clamping force acting in opposition to the restoring force of the deflected tether beams. The clamping forces needed to maintain various displacements were measured and used to determine static friction.

An improvement on this technique has recently been made by use of a scratch drive, or inchworm, actuator [14]. The device is shown schematically in Figure 5.4a, along with the sequence of clamping voltages required to create directional motion in Figure 5.4b. As can be seen, an electrostatically actuated plate of length L_p spans three frictional clamps of length L_c. At time zero, the voltage on the leading clamp is applied. Then the voltage on the middle clamp is applied, creating an out-of-plane deflection of amplitude A. The electrostatic force on the trailing clamp is then applied, followed by the release of the voltage on the leading clamp and then the voltage on the middle clamp. In this manner, the plate can travel in the chosen direction. At low force levels beam stretching is ignored and the step size, Δ, is given by [14]

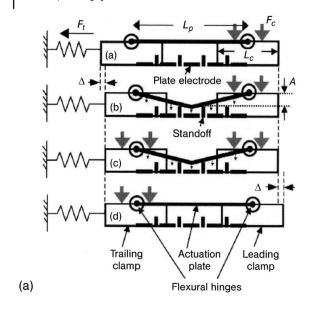

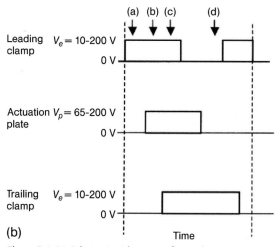

Figure 5.4 (a) Schematic side views of an inchworm actuator and (b) the sequence of clamping voltages required to create directional motion. Four sequential time steps are shown in each part [14].

$$\Delta = 2A^2/L_p \qquad (8)$$

For typical dimensions ($A = 3.1\,\mu m$ and $L_p = 500\,\mu m$), $\Delta = 38\,nm$. The scratch drive can travel an unlimited distance with a maximum force of $1.3\,mN$. (By contrast, the comb-drive shown in Figure 5.1 is only capable of $3\,\mu N$ at $200\,V$, for a polysilicon thickness of $3\,\mu m$.) Hence the scratch drive is capable of measuring friction over a much larger range of forces and displacements than the comb-drive actuated technique.

5.5
Wear

For MEMS devices that contain contacting parts, wear is a serious reliability concern. Wear effects have been characterized on-chip in the past simply by operating the devices over extended time periods and then investigating the morphology of the parts and surfaces using electron microscopy. Harmonic side-drive motors (wobble motors) [15] have been investigated, and also pin joints and gear hubs [16, 17]. In all cases, wear is a very real problem. Both erosion and debris formation are observed. Figure 5.5 shows an example of the erosion and wear debris that can accumulate and cause failure of a rotating device [16]. Environmental effects such as humidity and surface treatments such as fluorinated silane chains have been investigated.

5.6
Fracture Toughness and Strength

The fracture toughness, K_{Ic}, of a material is a constant for that material, regardless of microstructure [18]. However, measured fracture strengths can vary

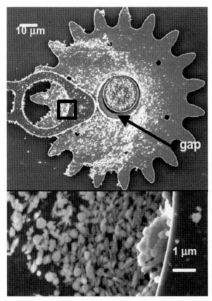

Figure 5.5 Scanning electron micrographs showing wear debris at a gear hub and pin joint after 600 000 cycles [16].

significantly, depending on the processing and surface finish of the components. For a brittle material, K_{Ic} is related to the measured fracture strength, σ_{crit}, by

$$K_{Ic} = c\sigma_{crit}(\pi a)^{1/2} \tag{9}$$

where a is the crack-initiating flaw size and c is a constant of order unity. The value of c depends on the exact size, shape and orientation of the flaw; for a semi-circular flaw, c is equal to 0.71 [19]. Therefore, any differences in the reported fracture strength of polysilicon will be due to changes in a. Typical MEMS structures are fabricated from chemical vapor deposited films. Therefore, the top and bottom surfaces are very smooth. The sidewalls, on the other hand, could contain significant roughnesses, due to the plasma etching or other etch techniques used to create the device geometries. As a result, measured fracture strengths can vary. For example, amorphous silicon structures with very smooth sidewalls had measured bend strengths about twice those of polysilicon structures with much rougher sidewalls [18].

Two types of on-chip techniques have been developed for measuring fracture toughness and fracture strength, either using passive devices that take advantage of residual stresses to generate the necessary forces or active devices that use electrostatic actuators to generate forces.

5.6.1
Passive Measurements of Fracture Toughness, Fracture Strength and Static Stress Corrosion

Figure 5.6 shows (a) a schematic top view, (b) a schematic side view and (c) scanning electron micrographs of a passive device for measuring fracture toughness [20]. The device consists of a fixed–fixed beam with an atomically sharp pre-crack perpendicular to the beam length. During fabrication, the pre-crack is introduced into the beam after etching of the beam, but before release. A Vickers microindent, with a load of 1 kg, is placed on the substrate near the beam. The high load causes radial cracks to form at the corners of the Vickers indenter tip, one of which travels into the polysilicon beam. Due to the stochastic nature of the radial cracks that form during indentation, the length of the pre-crack will vary somewhat. Therefore, a large number of identical beams can be fabricated with varying pre-crack lengths.

The residual tensile stress of the polysilicon film is measured, using one of the rotating strain gauges discussed in a previous section. Then, FEA is used to calculate the stress intensity factor, K, at the pre-crack tip as a function of the pre-crack length, when the device is released. If K is greater than the fracture toughness, K_{Ic}, the crack will propagate through the beam upon release. By accurately measuring the pre-crack lengths before release and then determining those lengths that propagate and those that do not, upper and lower bounds are established for K_{Ic}. This technique has been used for both polysilicon [20] and polycrystalline SiC [21, 22].

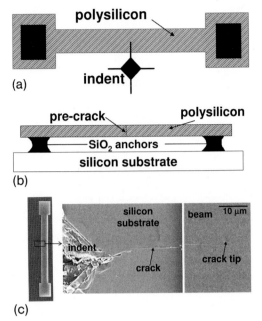

Figure 5.6 (a) Schematic top view, (b) schematic side view
and (c) scanning electron micrographs of a fixed–fixed
polysilicon beam containing an indent-produced atomically
sharp pre-crack. Upon release, the tensile stress in the beam
generates a stress intensity factor at the pre-crack tip [20].

Figure 5.7 shows a very similar device [23]. It also consists of fixed–fixed beams, but instead of an atomically sharp pre-crack, each beam contains a notch of varying length. Again, the residual tensile stress of the polysilicon is measured and the stress at the inside of each notch root upon release is calculated using FEA. For any notch longer than a critical length, the beam will break upon release. This device gives upper and lower bounds on the fracture strength of the material. Similarly, fixed–fixed beams can be fabricated that contain a region of decreased width in the center of the beam [24]. By varying the ratio of the reduced-width section to the entire beam length, the stresses in the reduced-width section can be changed. Again, fabricating a large number of these beams will give upper and lower bounds for the material's strength.

Since the loads are produced by the residual tensile stresses in the material, which will remain constant indefinitely at room temperature, the devices shown in Figures 5.6 and 5.7 are ideal for investigating static stress corrosion. The devices that did not fracture upon release are known to contain stresses and stress intensity factors just below the critical values. These devices can be placed in an environment of interest, such as high humidity, and the beams can be checked periodically for any time-dependent failures. No static stress corrosion was observed for polysilicon or polycrystalline SiC [20–23]. However, when the polysilicon had

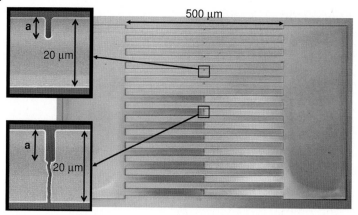

Figure 5.7 Scanning electron micrographs showing a top view of a set of fixed–fixed polysilicon beams containing notches of various lengths. After release, the tensile stress in the beams generated stress concentrations at the notch roots, such that the beams with notches longer than a critical length fractured [23].

a thin, thermally grown oxide surface layer, it was susceptible to static stress corrosion [23].

5.6.2
Active Measurements of Fracture Toughness and Fracture Strength

Instead of using residual tensile stresses to generate forces and stresses in the micromachined devices, actuators can be fabricated connected to the test specimens. The advantages are that the forces can be varied over a large range and are not fixed, as is the case for residual stresses. However, the main disadvantage is that the device design becomes much more complex. On-chip microactuators have been demonstrated for fracture strength measurements using: electrostatic comb-drives [25–27], electrostatic parallel plates [1], scratch drives [28] and thermal expansion actuators [29]. The microactuators and test specimens are fabricated simultaneously, using the same structural materials. Since they are made from the same materials using the same patterns, there are no issues with attachment or alignment.

Figure 5.8a shows a large electrostatic comb-drive actuator attached to a notched fracture strength specimen [30]. This specimen is shown in greater magnification in Figure 5.8b [18] and it consists essentially of a long, rigid cantilever beam anchored at one end, with a notch. When a voltage is applied to the comb-drive actuator, the device pulls down on the non-anchored end of the fracture strength specimen, creating a stress concentration at the notch root. As the voltage is increased, the force generated by the comb-drive increases, pulling the specimen

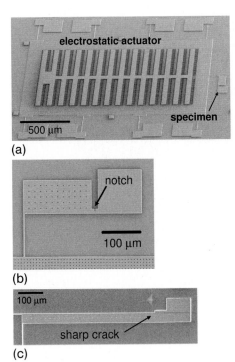

Figure 5.8 Scanning electron micrographs of (a) a large polysilicon comb-drive microactuator attached to a fracture specimen, (b) a higher magnification image showing the notched fracture specimen and (c) a fracture specimen with a sharp pre-crack, instead of a notch [18, 30].

further and increasing the stress at the notch root, until fracture occurs. The displacement of the actuator is monitored and FEA is used to translate the displacement of the actuator at the point of fracture to the stress at the notch root at that point, which defines the fracture strength, σ_{crit}.

The same comb-drive actuator can be fabricated with a fracture toughness specimen, shown in Figure 5.8c [18]. In this case, the notch seen in Figure 5.8b is replaced with an atomically sharp crack, which is created in the same manner as described for Figure 5.6. For this specimen, the movement of the actuator instills an increasing stress intensity factor at the crack tip, until the fracture toughness is reached and the specimen breaks catastrophically. This device has been fabricated from polysilicon with widely varying microstructures and demonstrates the microstructural independence of fracture toughness [18].

Figure 5.9 shows a parallel-plate electrostatic actuator, attached to a tensile specimen beam [1]. Parallel-plate actuators are capable of generating higher forces than comb-drive actuators, but the displacements generally are much smaller. Also, since the forces generated by the parallel plates increase sharply as the plate sepa-

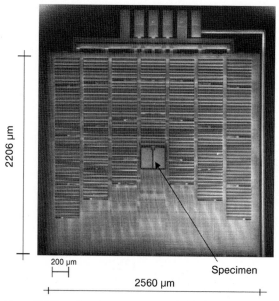

2206 µm

200 µm

Specimen

2560 µm

Figure 5.9 Scanning electron micrograph of a large polysilicon parallel-plate microactuator attached to a tensile beam fracture specimen [1].

ration decreases, the actuators are susceptible to pull-in, where the two plates abruptly snap together once a critical separation distance is reached. When the tensile beam is notched, sufficient forces are generated to cause fracture and determine fracture strength [1].

Two smaller actuators that have been used to study fracture strength in polysilicon are the scratch drive actuator, shown schematically in Figure 5.10 [28], and the thermal expansion actuator, shown in Figure 5.11 [29]. As described for Figure 5.4 the scratch drive operates by cycling the voltage between the device and the substrate. Since the separation between the two can be very small, the capacitance and thus the forces generated can be very large, even for relatively small devices. This device has been used to measure the tensile strength of polysilicon [28], but its difficulties lie in calibrating and controlling the forces applied to the tensile specimen.

The thermal expansion actuator in Figure 5.11 is also capable of generating high forces with typical lengths of 200–300 µm [29]. As the voltage is applied between the two outer anchors, the current passing through the polysilicon device causes a temperature increase due to Joule heating and the heated portion of the device will expand, creating tensile stresses in the fracture beam specimen. Here again, the difficulty is in calibrating the forces generated by the actuator. The temperatures throughout the device must be accurately monitored or modeled.

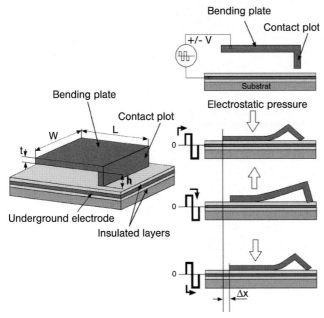

Figure 5.10 Schematic images showing the progressive motion of a scratch-drive microactuator [28].

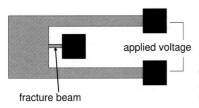

fracture beam

applied voltage

Figure 5.11 Schematic top view of a thermal expansion actuator. The solid features are anchors. Heating the device causes a tensile stress in the fracture beam.

5.7
Fatigue Measurements

Since polysilicon is a brittle material and displays essentially no dislocation activity at room temperature, it would not be expected to be susceptible to fatigue effects at room temperature. For bulk silicon, fatigue has not been reported. However, cyclic stress-induced failures are well documented for polysilicon MEMS devices.

By using a.c. instead of d.c. voltages, the microactuators shown in Figures 5.8 and 5.11 have been used to investigate the fatigue properties of the polysilicon structural materials. A collection of fatigue data for polysilicon is shown in Figure 5.12 [31]. The consistency of the data from several researchers shows that the life-

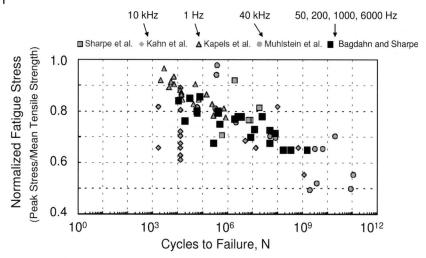

Figure 5.12 Previously published data for long-term fatigue of polysilicon MEMS specimens [31].

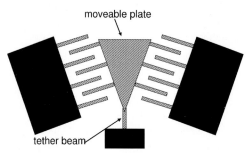

Figure 5.13 Schematic top view of an electrostatically comb-driven tilting fatigue device. The solid features are anchors.

time under high-cycle fatigue depends only on the number of cycles, not on the total time or the frequency of the test, for testing frequencies ranging from 1 Hz–40 kHz. This implies that dynamic fatigue depends only on the applied stresses and it must be mechanical in origin and not due to time-dependent environmental effects such as stress corrosion, oxidation or other chemical reactions.

In fact, dynamic fatigue of polysilicon has also been recently reported for tests in vacuum [32]. The fatigue lifetimes increased much more quickly for decreasing stresses than seen for the data in Figure 5.12, but the polysilicon is still susceptible to fatigue in vacuum environments. This implies that an air ambient exacerbates the fatigue effects in polysilicon, but it is not necessary for fatigue. The vacuum tests were performed using a tilting comb-drive device like that shown schematically in Figure 5.13. There are many fewer comb fingers than in the device shown in Figure 5.8 and normally sufficient force to cause fracture would not be possible.

However, the device takes advantage of the resonant frequency and the very low damping that occurs in vacuum ambients. When the a.c. drive voltage applied to the comb-drive actuator is equal to the resonant frequency of the device, large vibrational amplitudes can be achieved even for very modest forces.

Some of the data in Figure 5.12 were generated using a device similar to that shown in Figure 5.13 [33, 34] and similar results were also obtained for single-crystal silicon [35]. For those tests, the origin of the fatigue was postulated to be caused by the growth of a surface oxide in the areas of high stress and the subsequent static stress corrosion of the oxide (unlike polysilicon, silicon dioxide is susceptible to static stress corrosion [36]). However, these devices contained gold contact pads. It has recently been demonstrated that during the release process in aqueous hydrofluoric acid, if polysilicon devices contain gold contact pads, galvanic corrosion of the outer surface of the polysilicon will take place [37–40]. This results in a layer of porous silicon on the surface of the polysilicon devices, which will readily oxidize on exposure to rinse water or air. As a result, those devices [33, 34] possessed a surface that consisted of porous silicon and a thick surface oxide and the results for fatigue of this material cannot necessarily be applied to fatigue of polysilicon. It must be noted, however, that many characteristics of the fatigue behavior, such as stress dependence and environment dependence, match those of polysilicon. Recently, the experiments were repeated for polysilicon devices without gold contact pads, and the results showed the same trends with thinner surface oxides and higher fatigue strengths [41].

5.7.1
Fatigue Under Varying Mean Stresses and Fatigue Amplitudes

The data in Figure 5.12 were all taken either for fully reversed loading (equal tensile and compressive stresses in the cycle) or for tension-zero cycling. The comb-drive microactuator shown in Figure 5.8 contains such a large number of comb fingers that it can cause fracture of the notched polysilicon specimens under a monotonically increasing d.c. voltage. Therefore, it can also be used to study fatigue of polysilicon with a varying mean stress, σ_m. Figure 5.14 shows a schematic of a fatigue experiment where the mean stress and the fatigue stress amplitude, $\Delta\sigma$, are varied independently [30]. Because it is desirable to investigate the effects of compressive fatigue stresses, in addition to tensile fatigue stresses, the protocol shown in Figure 5.14 was carried out. In this test, the specimens are first

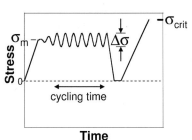

Figure 5.14 Schematic representations of the stresses seen at the notch of the specimens shown in Figure 5.8, during a fixed-time, constant $\Delta\sigma$ fatigue test [30].

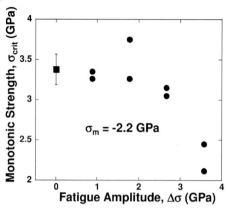

Figure 5.15 Results from constant $\Delta\sigma$ fatigue tests, where σ_m was -2.2 GPa for all tests. The monotonic strength, taken from specimens that saw no cycling, is shown as the solid square; the square marks the average strength and the error bars represent one standard deviation [30].

taken to the σ_m by use of a d.c. voltage. Then, a cyclic stress at a fixed $\Delta\sigma$ is gener-ated using an additional a.c. voltage at the resonance frequency of the device (about 10 kHz). The specimens are subjected to six million stress cycles and then the loads are released. After this, the specimen is subjected to a monotonically increas-ing load until fracture occurs and the fracture strength, σ_{crit}, is determined. The results for the fracture strength of the cycled specimens are then compared with the fracture strength of specimens that experienced no cyclic stresses.

Figure 5.15 shows results for constant $\Delta\sigma$ tests (Figure 5.14) where σ_m was fixed at -2.2 GPa and $\Delta\sigma$ was varied. The square symbol shows the fracture strength of specimens that saw no cyclic stresses and the error bar represents one standard deviation in the strength values. For this σ_m, cycling with small $\Delta\sigma$ does not affect the σ_{crit} measured after cyclic stressing, but cycling with large $\Delta\sigma$ leads to a decrease in σ_{crit}.

Figure 5.16 shows the results for constant $\Delta\sigma$ tests of polysilicon specimens where $\Delta\sigma$ was fixed at 2.0 GPa (±1.0 GPa) and σ_m was varied. Again, the square symbol represents the fracture strength of specimens that saw no cyclic stresses and the error bar represents one standard deviation in the strength values. For small tensile or compressive σ_m, σ_{crit} is unaffected by the relatively small $\Delta\sigma$ cycling. However, for large tensile or compressive σ_m, σ_{crit} is increased – the speci-mens have strengthened. In the experiments with the highest tensile σ_m, 2.0 GPa, the maximum tensile stress seen at the notch root during cycling was 3.0 GPa. This exceeds the average monotonic bend strength, 2.7 GPa, although it falls within one standard deviation, 0.4 GPa. As shown in Figure 5.14, once the σ_m is applied, the cyclic stresses are ramped up to the desired $\Delta\sigma$. None of the five specimens tested under these conditions broke when the cyclic stresses were ramped up to a maximum stress in the cycle of 3.0 GPa. Statistically, it is highly

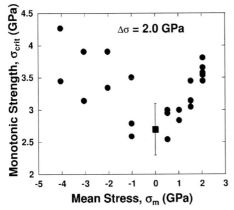

Figure 5.16 Results from constant $\Delta\sigma$ fatigue tests, where $\Delta\sigma$ was 2.0 GPa for all tests. The monotonic strength, taken from specimens that saw no cycling, is shown as the solid square; the square marks the average strength and the error bars represent one standard deviation [30].

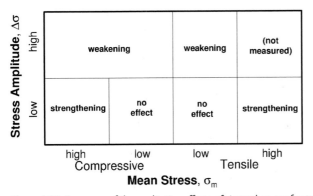

Figure 5.17 Summary of the qualitative effects of $\Delta\sigma$ and σ_m on fracture strength, σ_{crit} [30].

unlikely that all five specimens would have displayed σ_{crit} greater than 3.0 GPa without any cycling. Therefore, it is presumed that during the ramp up of the cyclic stress, which takes several seconds ($\sim 10^5$ cycles), enough strengthening occurred in the specimens to survive a maximum stress in the cycle of 3.0 GPa. The trends observed in these experiments are summarized in Figure 5.17 [30].

Weakening during compressive cycling has previously been reported for bulk ceramics [42]. In that study, it was assumed that the cyclic compressive stresses caused some irreversible damage in the vicinity of the stress concentration (at the notch root). This is also the likely origin for the weakening and fatigue susceptibility in polysilicon MEMS, although the exact mechanism of the irreversible damage remains unknown.

The origin of the strengthening is also not known definitively. As stated above, dislocation activity in polysilicon is believed not to occur at room temperature. One possible explanation is the creation of local compressive stresses at the notch root during cycling. These stresses could come about if the polysilicon experiences any grain boundary plasticity [30]. It is also important to note that polysilicon fatigue tests in vacuum also demonstrated strengthening effects [32]. In that study, all of the tests were done for fully reversed loading (σ_m equal to zero). The strengthening occurred for specimens that were cycled for a long time with a relatively low fatigue amplitude.

References

1 A. Corigliano, B. De Masi, A. Frangi, C. Come, A. Villa, M. Marchi, *J. Microelectromech. Syst.* **13** (2004) 200–219.

2 W. C. Tang, T.-C. H. Nguyen, R. T. Howe, *Sens. Actuators* **20** (1989) 25–32.

3 H. Kahn, M. A. Huff, A. H. Heuer, *MRS Symp. Proc.* **518** (1998) 33–38.

4 A. Kaushik, H. Kahn, A. H. Heuer, *J. Microelectromech. Syst.* **14** (2005) 359–267.

5 R. I. Pratt, G. C. Johnson, *MRS Symp. Proc.* **518** (1998) 15–20.

6 B. D. Jensen, M. P. de Boer, N. D. Masters, F. Bitsie, D. A. LaVan, *J. Microelectromech. Syst.* **10** (2001) 336–346.

7 M. S. Baker, M. P. de Boer, N. F. Smith, L. K. Warne, M. B. Sinclair, *J. Microelectromech. Syst.* **11** (2002) 743–753.

8 H. Guckel, D. Burns, C. Rutigliano, E. Lovell, B. Choi, *J. Micromech. Microeng.* **2** (1992) 86–95.

9 T. Kramer, O. Paul, *Sens. Actuators A* **92** (2001) 292–298.

10 B. P. van Drieenhuizen, J. F. L. Goosen, P. J. French, R. F. Wolffenbuttel, *Sens. Actuators A* **37–38** (1993) 756–765.

11 F. Ericson, S. Greek, J. Soderkvist, J.-A. Schweitz, *J. Micromech. Microeng.* **7** (1997) 30–36.

12 W. R. Ashurst, C. Yau, C. Carraro, R. Maboudian, M. T. Dugger, *J. Microelectromech. Syst.* **10** (2001) 41–49.

13 M. G. Lim, J. C. Chang, D. P. Schultz, R. T. Howe, R. M. White, *Proc. IEEE Micro Electro Mechanical Systems Workshop* (1990) 82–88.

14 M. P. de Boer, D. L. Luck, W. R. Ashurst, R. Maboudian, A. D. Corwin, J. A. Walraven, J. M. Redmond, *J. Microelectromech. Syst.* **13** (2004) 63–74.

15 M. Mehregany, S. D. Senturia, J. H. Lang, *Proc. IEEE Solid-State Sensor and Actuator Workshop* (1990) 17–22.

16 D. M. Tanner, J. A. Walraven, L. W. Irwin, M. T. Dugger, N. F. Smith, W. P. Eaton, W. M. Miller, S. L. Miller, *Proc. IEEE Int. Reliability Phys. Symp.* (1999) 189–197.

17 J. A. Walraven, T. J. Headley, A. N. Campbell, D. M. Tanner, Proc. SPIE 3880 (1999) 30–39.

18 R. Ballarini, H. Kahn, N. Tayebi, A. H. Heuer, *Mechanical Properties of Structural Films*, ASTM STP 1413, 37–51, ASTM, Philadelphia, PA 2001.

19 I. S. Raju, J. C. Newman, Jr., *Eng. Fracture Mech.* **11** (1979) 817–829.

20 H. Kahn, R. Ballarini, J. J. Bellante, A. H. Heuer, *Science* **298** (2002) 1215–1218.

21 V. Hatty, H. Kahn, J. Trevino, M. Mehregany, C. A. Zorman, R. Ballarini, A. H. Heuer, *J. Appl. Phys.* **99** (2006) 013517–013517-5.

22 J. J. Bellante, H. Kahn, R. Ballarini, C. A. Zorman, M. Mehregany, A. H. Heuer, *Appl. Phys. Lett.* **86** (2005) 071920–071920-3.

23 H. Kahn, R. Ballarini, A. H. Heuer, *MRS Symp. Proc.* **741** (2003) J3.4.1–J3.4.6.

24 M. Biebl, H. von Philipsborn, *Proc. Int. Conf. Solid-State Sensors and Actuators, Transducers* 95 (1995) 72–75.

25 H. Kahn, R. Ballarini, R. L. Mullen, A. H. Heuer, *Proc. R. Soc. London A* **455** (1990) 3807–3823.

26 S.-H. Lee, J. S. Kim, J. W. Evans, D. Son, Y. E. Pak, J. U. Jeon, D. Kwon, *Microsyst. Technol.* **7** (2001) 91–98.

27 S.-H. Lee, J. W. Evans, Y. E. Pak, J. U. Jeon, D. Kwon, *Thin Solid Films* **408** (2002) 223–229.

28 P. Minotti, P. Le Moal, E. Joseph, G. Bourbon, *Jpn. J. Appl. Phys.* **40** (2001) L120–L122.

29 H. Kapels, R. Aigner, J. Binder, *IEEE Trans. Electron Devices* **47** (2000) 1522–1528.

30 H. Kahn, L. Chen, R. Ballarini, A. H. Heuer, *Acta Mater.* **54** (2006) 1597–1606.

31 J. Bagdahn, W. N. Sharpe, Jr., *Sens. Actuators A* **103** (2003) 9–15.

32 R. Boroch, presented at "Mechanical Reliability of Silicon MEMS – Recent Progress and Further Requirements," 27–28 February 2006, Halle, Germany.

33 C. L. Muhlstein, E. A. Stach, R. O. Ritchie, *Acta Mater.* **50** (2002) 3579–3595.

34 C. L. Muhlstein, S. B. Brown, R. O. Ritchie, *Sens. Actuators A* **94** (2001) 177–188.

35 O. N. Pierron, C. L. Muhlstein, *J. Microelectromech. Syst.* **15** (2006) 111–119.

36 T. A. Michalske, S. W. Freiman, *J. Am. Ceram. Soc.* **66** (1983) 284–288.

37 M. Huh, Y. Yu, H. Kahn, J. Payer, A. H. Heuer, *J Electrochem. Soc.* in press.

38 H. Kahn, C. Deeb, I. Chasiotis, A. H. Heuer, *J. Microelectromech. Syst.* **14** (2005) 914–923.

39 D. C. Miller, K. Gall, C. R. Stoldt, *Electrochem. Solid-State Lett.* **8** (2005) G223–G226.

40 O. N. Pierron, D. D. Macdonald, C. L. Muhlstein, *Appl. Phys. Lett.* **86** (2005) 211919–211919-3.

41 D. H. Alsem, R. Timmerman, B. L. Boyce, E. A. Stach, J. Th. M. De Hosson, R. O. Ritchie, *J. Appl. Phys.* **101** (2007) 013515.

42 L. Ewart, S. Suresh, *J. Mater. Sci.* **22** (1987) 1173–1192.

6
Reliability of a Capacitive Pressure Sensor

Fumihiko Sato, Hideaki Watanabe, Sho Sasaki, Corporate Research and Development Headquarters, OMRON Corporation, Kizu, Kyoto, Japan

Abstract

A new silicon capacitive pressure sensor with center-clamped diaphragm is presented. The sensor has a silicon–glass structure and is fabricated by a batch fabrication process. Since a deformed diaphragm has a donut shape and parallel-like displacement is realized and, therefore, better linearity of 0.7%, which is half of that of the conventional flat diaphragm sensor, is obtained. It was clarified both analytically and experimentally that the capacitive pressure sensor with center-clamped diaphragm is advantageous in terms of linearity. Furthermore, the design and fabrication process for high reliability, the mechanism of capacitive change, which depends on humidity, the resonance phenomenon of the diaphragm and the surge of static electricity are described.

Keywords

capacitance; pressure sensor; linearity; reliability; diaphragm; micromachining

Reliability of MEMS.
Edited by O. Tabata, T. Tsuchiya
Copyright © 2013 WILEY-VCH Verlag GmbH & Co. KGaA, Weinheim
ISBN: 978-3-527-33501-5

6.1
Introduction

Micromachined pressure sensors have been developed because of their small size, high performance, high reliability and low cost. These sensors are required in industrial, factory, medical and automotive fields. They are divided into two categories: capacitive and piezoresistive sensors. Capacitive sensors can easily obtain frequency output with conventional CR oscillators. In addition, they have distinct advantages such as high sensitivity, better temperature performance, low power consumption and easy interfacing with microcomputers [1, 2] compared with their piezoresistive counterparts. On the other hand, they exhibit poor linearity because of diaphragm deformation that is far from parallel movement. In order to obtain better linearity, parallel displacement of the moveable electrode to the fixed electrode is desired. Several diaphragm structures have been reported for obtaining better linearity, such as diaphragms with stiffened mesa regions that produce displacement without bending in the moveable electrode [3–5]. These mesa structures offer better linearity than the flat diaphragm structures. However, these diaphragms are large and require precise process control for fabrication because of the complicated compensation mask pattern for anisotropic etching.

We have developed a new diaphragm structure that is circular in shape and clamped at both the center and the rim. It offers both better linearity and a simpler structure for easier fabrication. This chapter describes the structure, characteristics, mechanical reliability, electrical reliability and humidity characteristics of the capacitive pressure sensor with the new diaphragm.

6.2
Structure and Principle

The structure of the new capacitive pressure sensor with center-clamped diaphragm is shown in Figure 6.1. The sensor consists of silicon and glass layers. A ring-like fixed electrode for the detection of capacitance changes is formed on the bottom side of the top glass layer. The silicon layer has a thin membrane that is fabricated by anisotropic etching from the bottom surface. A circular shallow groove is formed on the upper side of the silicon membrane. A small silicon pillar stands at the center of the groove. The silicon layer is electrostatically bonded to the glass layer at the rim of the groove and at the center pillar on the membrane, so that the diaphragm that operates as the moveable electrode is ring shaped. Figure 6.2 shows a scheme of this diaphragm. The fixed electrode is connected to an aluminum pad on the silicon layer through bumps. The gap between the electrodes is spatially connected to the atmosphere through the feedthrough groove for measuring differential pressure.

Figure 6.3 shows calculated results for the center-clamped diaphragm deformation obtained by the finite element method (FEM). The model is a ring-like thin plate clamped at the both the outer and inner rim. The shape of the deformed

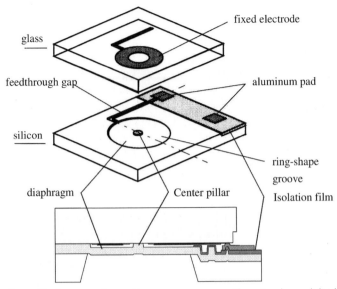

Figure 6.1 Structure of capacitive pressure sensor with center-clamped diaphragm.

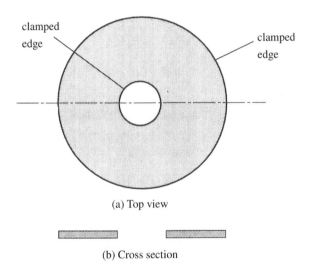

(a) Top view

(b) Cross section

Figure 6.2 Model of center-clamped diaphragm.

diaphragm becomes donut-like. The region of maximum deflection of the diaphragm is ring shaped and so parallel-like displacement of the moveable electrode to the fixed electrode can be achieved as opposed to the case with the conventional flat plate diaphragm. Hence this sensor obtains better linearity than a flat-plate diaphragm sensor. Because the diaphragm of this sensor can be fabricated from

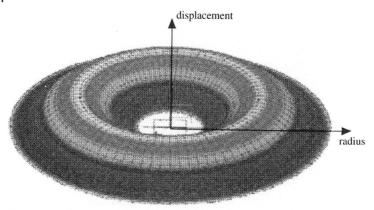

Figure 6.3 Deformation of center-clamped diaphragm calculated by finite element method.

a flat silicon membrane, the fabrication process for the diaphragm does not need complicated compensation pattern technology to form the mesa structure. Therefore, the sensor can be fabricated by the same process that is used for the conventional simple flat diaphragm sensor.

6.3
Finite Element Analysis

The mechanical behavior of the center-clamped diaphragm was evaluated by FEM. IDEAS was used as the FEM program. One-quarter of the ring-like plate was modeled by a 2D shell. The modeled diaphragms had radii of 0.5–1 mm, a thickness of 2–10 μm and center pillar radii of 10–300 μm. The mechanical properties of silicon, a Young's modulus of 190 GPa and a Poisson's ratio of 0.28 [6] were used.

First, the deflections of the several models were calculated. Figure 6.4 shows the deflection distribution against the distance from the center of the diaphragm, calculated for the model with a diaphragm radius of 0.75 mm, a thickness of 6 μm and a center pillar radius of 150 μm at an applied pressure of 1 kPa. From the results, the deflection shows a maximum value at a point 420 μm from the center of the diaphragm. The point is located at a distance of 45% of the moveable ring-like region width from the rim of the center pillar, which is equal to the difference between the diaphragm radius and the center pillar radius. Figure 6.5 shows the maximum deflection against the width of the moveable ring for several calculated models with a diaphragm thickness of 6 μm at an applied pressure of 1 kPa. Figure 6.6 shows the maximum deflection against the diaphragm thickness for models with a diaphragm radius of 0.75 mm and a center pillar radius of 150 μm at an applied pressure of 1 kPa. From these results, the maximum deflection of the center-clamped diaphragm can be expressed as

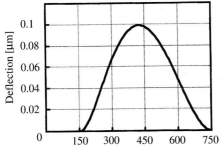

Figure 6.4 Diaphragm deflection as a function of the distance from center of the diaphragm at an applied pressure of 1 kPa. A diaphragm radius of 0.75 mm, a thickness of 6 μm and a center pillar radius of 0.15 mm were used.

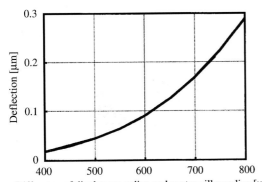

Figure 6.5 Maximum deflection of diaphragm against difference between diaphragm radius and center pillar radius. A thickness of 6 μm and a center pillar radius of 150 μm were used.

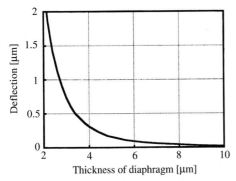

Figure 6.6 Maximum deflection of diaphragm against thickness. A diaphragm radius of 0.75 mm and a center pillar radius of 0.15 mm at an applied pressure of 1 kPa were used.

$$\omega_{max} = 1.52 \times 10^{-13} \times \frac{(a-b)^4 P}{t^3} \tag{1}$$

wherein a, t, b and P are the diaphragm radius, the diaphragm thickness, the center pillar radius and the applied pressure, respectively. The maximum deflection is found to be proportional to $(a-b)^4$ and inversely proportional to t^3. This result is similar to that for flat plate diaphragms with clamped edges.

The stress was also calculated by FEM. Models with a diaphragm radius of 0.75 mm and a center pillar radii of 10–300 μm were used. The thickness of the diaphragms was determined to have the same deflection of 0.1 μm at an applied pressure of 1 kPa. The maximum Von Mises stress was induced at the rim of the center pillar of the diaphragm. Figure 6.7 shows the maximum stress against the radius of the center pillar at an applied pressure of 10 kPa. From Figure 6.7, it is seen that the induced stress can be decreased by increasing the center pillar radius and saturates over the radius of 150 μm. Since the survival pressure of the diaphragm is equal to the applied pressure that induces a maximum stress larger than the fracture stress of single-crystal silicon, 7 GPa [6], the durability of the diaphragm can be designed by the dimensions of the center pillar.

The electrical properties of the capacitive pressure sensor having a center-clamped diaphragm were calculated from the FEM analysis results. The sensor models were designed for an operating pressure range of 0–1 kPa. The sensor had a diaphragm radius of 0.75 mm, a center pillar radius of 150 μm and a diaphragm thickness of 2–10 μm. The center pillar radius was determined to minimize the induced stress of the diaphragm. The sensitivity was estimated from the ratio of the capacitance change by an applied pressure between 1 and 0 kPa to the initial capacitance at 0 kPa. The non-linearity was estimated using the reciprocal of the capacitance, because the capacitance is converted to frequency, which is inversely proportional to the capacitance. The fixed electrode was originally designed for minimizing non-linearity. Figure 6.8 shows the non-linearity as a function of the

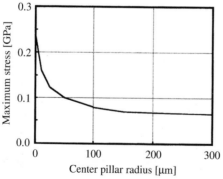

Figure 6.7 Maximum stress as a function of the center pillar radius. The model with a diaphragm radius of 0.75 mm and center pillar radii from 10 to 300 μm was used.

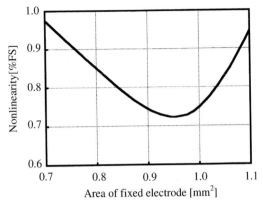

Figure 6.8 Non-linearity distribution on area of fixture electrode.

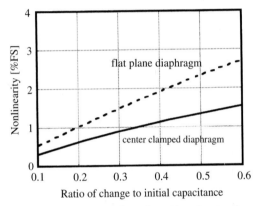

Figure 6.9 Comparison of non-linearities of center-clamped diaphragm and flat diaphragm sensor as a function of the ratio of changed to initial capacitance. The center-clamped diaphragm is represented by the solid line and the flat diaphragm by the broken line.

area of the fixed electrode. The non-linearity occurs because of the incomplete parallel displacement of the moveable electrode against the fixed electrode. Also, the parasitic capacitance causes non-linearity. Therefore, the area of the fixed electrode is limited by both the incomplete parallel displacement of the diaphragm and the parasitic capacitance. The non-linearities of the sensors were compared with that of the conventional flat circular plate diaphragm sensor. The flat-plate diaphragm had a radius of 0.75 mm and a thickness of 6–15 μm to produce the same ratio of change to initial capacitance from 0.1 to 0.6. The fixed electrode of the flat diaphragm sensor was circular and had the same area as the center-clamped diaphragm sensor. The non-linearities of both sensors are shown in the scale of the ratio of change to initial capacitance in Figure 6.9.

From the results, the non-linearity of the center-clamped diaphragm sensor is about half of that of the flat diaphragm sensor at all ratios of change to initial capacitance. Therefore, the new diaphragm is very efficient and improves the measurement accuracy via better linearity.

6.4
Fabrication Process

The fabrication process of the sensor is shown schematically in Figure 6.10 and the various steps are outlined below.

(a) A double-sided polished (100) silicon substrate (400 μm thick) is used. The trench for air intake is etched using tetramethylammonium hydroxide (TMAH) solution with an insulating oxide film as an etch mask. A bump for avoiding dust intrusion during the dicing and device operation is placed in the trench.

(b) A capacitance gap with center pillar and bumps for interconnections is formed using TMAH solution, etching through a thermally oxidized film again as a etch mask.

(c) Phosphorus is implanted and diffused to form an n-type region of suitable junction depth on the upper side of the silicon substrate. The membrane thickness of the pressure-sensitive diaphragm can be controlled precisely and inexpensively by using a p–n junction etch-stop technique. In this phase, the process condition to reduce the leak current of the p–n junction is important for precise etch-stop control, for example, suppression of crystal defects.

(d) Thermal oxide is grown and a silicon nitride film is deposited by low-pressure chemical vapor deposition (LP-CVD) on the bottom side of the substrate and patterned to form the anisotropic etching mask.

(e) An LP-CVD silicon dioxide film is deposited on the upper side of substrate and patterned to form the isolation pattern for the aluminum pad.

(f) LP-CVD silicon dioxide films are deposited on both sides of substrate and a contact hole is formed by etching the isolation film. A highly doped layer is formed on the surface of the contact hole by phosphorus deposition.

(g) Aluminum film is sputtered and patterned to form the pads for wire bonding and the bump that connects the pad on the silicon substrate to the fixed electrode on the glass substrate.

(h) Isolation film is patterned to form a 0.3-μm gap as air inlet.

(i) A 500-μm thick glass substrate with a thermal expansion coefficient close to that of silicon is used for better temperature

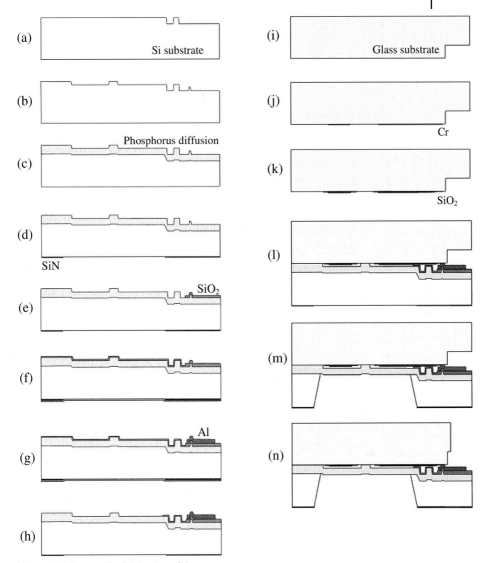

Figure 6.10 Process for fabrication of the sensor.

properties. A groove is etched with HF solution to expose the bonding pad.

(j) Chromium is sputtered and patterned to form a fixed electrode on the glass substrate.

(k) A silicon dioxide isolation film is sputtered on the chromium to improve the amount of static electricity and to avoid welding between the fixed and moveable electrodes.

(l) The glass substrate is anodically bonded to the silicon substrate. If the bonding strength is not good enough, a defect
would occur under the high temperature and humidity
condition like 358k and 85%, because pressure is applied to
the diaphragm, and that leads to lower sensitivity. Therefore,
the substrate is bonded at 613K, 800V in nitrogen ambient
in order to obtain enough strength as the condition at low
temperature in a short amount of time. In this phase, the
interconnection is formed by pressure bonding between
the chromium on the glass substrate and the aluminum
on the silicon substrate. Contact is made by five circular
bumps 10 μm in diameter that have a sufficient aluminum
membrane thickness to collapse. The high reliability of the
interconnection was proven by thermal cycle testing and
heat shock testing.

(m) The silicon substrate is etched from the bottom side with
45 wt.% KOH solution at 343 K to form a silicon membrane
which is larger than the ring-shaped gap in it. Using the p–n
junction etch-stop technique, the membrane thickness can
be precisely controlled.

(n) The top side of the glass substrate is diced half way along the
groove on the bottom side of the glass substrate to expose
the bonding pad and is diced fully to the individual sensor
chip.

6.5
Fundamental Characteristics

A photograph of the capacitive pressure sensor chip with center-clamped diaphragm is shown in Figure 6.11. The sensor is 3 mm square with a diaphragm

Figure 6.11 Photograph of capacitive pressure sensor with center-clamped diaphragm.

diameter of 1.5 mm, a thickness of 6 µm and a center pillar diameter of 300 µm. A flat diaphragm sensor with a diaphragm diameter of 1.5 mm and a thickness of 9.8 µm was prepared for comparison with the characteristics of the center-clamped diaphragm sensor. The characteristics of the sensors were measured with an impedance meter at a frequency of 1 MHz. The measurement capacitance of the sensors was 11.5 pF at an applied pressure of 0 kPa and 14.2 pF and the full operating range of 1 kPa. The ratio of changed to initial capacitance was 0.23. Figure 6.12 shows the capacitance change of the center-clamped sensor as a function of the applied pressure. The capacitance change is inversely proportional to the pressure. The non-linearities of both sensors are shown in Figure 6.13. The non-linearity of the center-clamped diaphragm sensor was 0.7% full-scale (FS), which was half of that of the flat diaphragm sensor (1.4% FS). The experimental values are almost equivalent to the calculated results and the ratio of changed to initial capacitance of 0.23, shown in Figure 6.9.

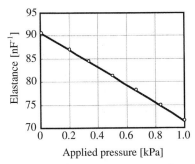

Figure 6.12 Electric capacitance as a function of applied pressure.

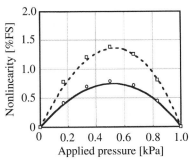

Figure 6.13 Non-linearities of both center-clamped diaphragm sensor and flat diaphragm sensor as a function of applied pressure. Measurement values of center-clamped diaphragm are plotted with circles and flat diaphragm as squares and the calculated results are represented by the solid curve for the center-clamped diaphragm and the broken curve for the flat diaphragm.

The thermal characteristics in the temperature range 303–343 K were measured. The thermal offset drift was 130 ppm K^{-1} and the sensitivity shift was 240 ppm K^{-1}. These results were almost the same as those for the flat diaphragm sensor.

From the comparison of the non-linearities, it is seen that the non-linearity of the diaphragm type of capacitive pressure sensor is reduced by the center-clamped diaphragm structure. It is believed that a parallel-like displacement of the moveable electrode and an optimized design of the fixed electrode have been achieved.

6.6
Mechanical Reliability

Single-crystal silicon and polysilicon are well known as moveable electrode materials of high mechanical reliability in the MEMS field, because the fatigue which can be observed in metallic material is basically not observed in brittle materials. In this section, characteristic problems of mechanical reliability that depend on the device structure of the capacitive pressure sensor are described.

The gap between the fixed and moveable electrodes of the capacitive sensor is generally designed to be as small as possible in order to deliver miniaturization and high sensitivity. The capacitive pressure sensor has closely facing fixed and moveable electrodes and the gap is about 1 μm. Therefore, the stiction of the diaphragm by liquid intrusion and agglomeration and movement inhibition by dust intrusion cause the reliability to deteriorate under the pressure cycles.

6.6.1
Stiction

The prevention of stiction is an inevitable problem for high reliability of MEMS. Surface hydrophobization by coating with a fluorinated thin film, minimization of the contact area by using microdimples and so forth are adopted.

A capacitive pressure sensor may fail owing to stiction, in which the moveable electrode on the silicon diaphragm adheres to the fixed electrode and cannot work correctly.

In general, stictions are caused by liquid bridge formation force (meniscus force), van der Waals force and electrostatic force. However, in the case of a capacitive sensor that introduces atmospheric pressure into the gap, liquid bridge formation force is the main cause.

Figure 6.14 shows an example of stiction that occurs in mechanical reliability testing such as a pressure proof test, because liquid in the gap agglomerates and generates a bridge formation force in excess of the restorative force for detachment. Therefore, the process needs to be managed so as to eliminate the residual liquid contamination such as rinse liquid by thorough cleaning and drying.

Alternatively, from the viewpoint of device design, if the restorative force were larger than the adherence force, stiction would never occur. Considering the rela-

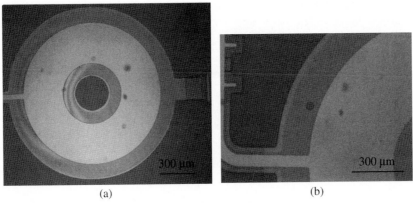

(a) (b)

Figure 6.14 Example of stiction occurring in the mechanical reliability test, showing (a) the whole picture of the stiction and (b) the liquid-induced stiction.

tion between the sensitivity and stiction, a design aimed at increasing the restorative force by improving the spring constant of the diaphragm is important.

6.6.2
Dust

If dust intrudes into the gap of a capacitive pressure sensor that introduces atmospheric pressure into the gap, this prevents the diaphragm from deforming and leads to linearity errors. Countermeasures should be taken to prevent dust intrusion during fabrication and operation.

The cleaning process is implemented as a dust intrusion countermeasure during fabrication. However, if dust intrudes into the gap during fabrication, it would displace during repeated pressure and vibration tests and may lead to linearity errors. A fringe pattern inspection that can find and eliminate devices afflicted with dust intrusion is adopted to avoid this reliability defect. Figure 6.15 shows normal and dust-intruded chips This sensor has a circular diaphragm and, if it is pressed until the diaphragm contacts the fixed electrode, a concentric fringe pattern can be observed through the glass substrate because of the gap between the fixed electrode and diaphragm. When dust exists in the gap, regardless of its size, a disorder appears in the fringe pattern. The symmetry of the fringe pattern is inspected automatically and a device that has dust on the inside can be easily distinguished and eliminated. The dust trap is formed by a narrow stricture and dent as a dust intrusion countermeasure during working. Figure 6.16 depicts its structure. This sensor is designed to have a 0.3-μm gap between the fixed electrode and diaphragm when the maximum detection pressure is impressed. Dust particles larger than 0.3 μm would not enter and hence not lead linearity errors.

The regularly uses pressure cycle test (0–3.4 kPa, 4×10^6 cycles), the maximum pressure range test (0–9.8 kPa, 2×10^6 cycles) and the characteristic change test

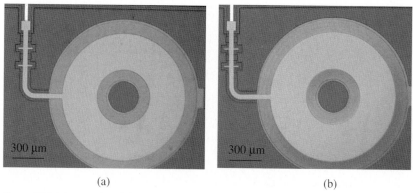

(a)   (b)

Figure 6.15 (a) Normal and (b) dust-intruded chips.

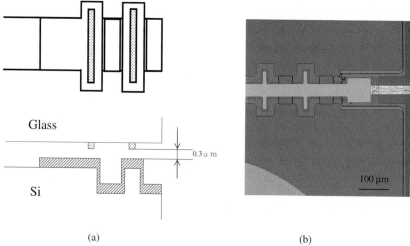

(a)   (b)

Figure 6.16 (a) The dust-trap structure and (b) picture of the upper part of (a). The bottom part of (a) is the cross-section of the upper part.

(3 V DC, 2.9 kPa at 363 K) were examined to evaluate mechanical reliability. Figure 6.17 shows the results of the tests. The fluctuation of the capacitance value is less than ±3%, which is excellent.

6.7
Humidity Characteristics

In the case of a differential capacitive pressure sensor that measures the difference in pressure between two ports, the variation of output capacitance of the sensor

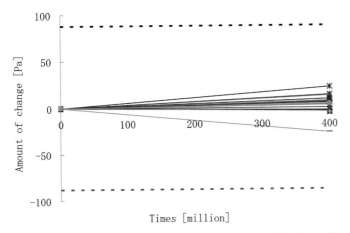

(a) Output drift at P = 0 kPa after cyclic pressure application of 0 – 3.4kPa.

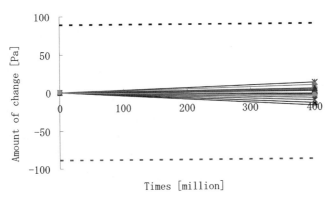

(b) Output drift at P = 4.9 kPa after cyclic pressure application of 0 – 3.4kPa.

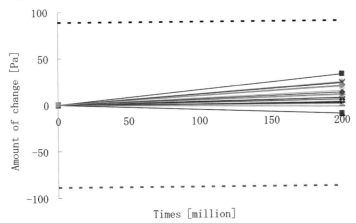

(c) Output drift at P = 0 kPa after cyclic pressure application of 0 – 9.8 kPa.

Figure 6.17 The repeat duration test for reliability. Broken lines show ±3% of the full-scale fluctuation of the capacitance.

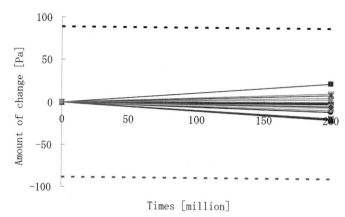

(d) Output drift at P = 4.9 kPa after cyclic pressure application of 0 – 9.8 kPa.

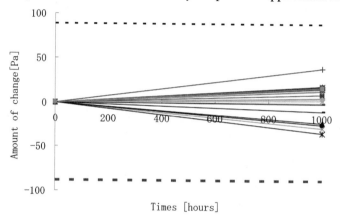

(e) Output drift at P = 0 kPa after static pressure application of 2.9 kPa at 365K.

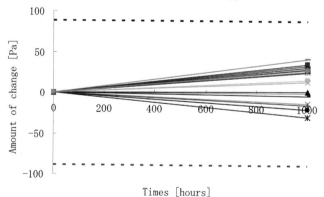

(f) Output drift at P = 4.9 kPa after static pressure application of 2.9 kPa at 365K.

Figure 6.17 *Continued*

by atmospheric humidity is large because the humid air may intrude directly into the gap between the fixed and moveable electrodes. In this section, the mechanism of capacitance variation caused by humidity is described.

The sensor chip was die bonded to a plastic-stem package with silicone adhesive and wire bonded to the stem terminals. The output of the sensor was measured in a constant-temperature and -humidity oven (313 K, 95% RH) and subsequently compared and evaluated against the output of the sensor at 313 K, 50% RH. The humidity effects of the sensor had a tendency such that the output capacitance increased as the humidity increased. A value of 11 pF was observed at 313 K, 50% RH, but it increased by approximately 0.1 pF in a 313 K, 95% RH high-humidity atmosphere. Because the glass surface around the sensor electrode (chromium) has a lower surface resistance when the humidity increases and the electric charge can be spread easily from the sensor electrode to the glass surface, the sensor electrode appears to be larger. We interpreted the cause as being that the glass surface could absorb water molecules during anodic bonding. The reasons for this can be as follows:

1. The sensor electrode causes it, because the capacitance change without a diaphragm due to humidity is approximately 0.02 pF, which is very small.
2. It is highly unlikely that the leakage current that flows around the glass surface in the sensor gap between the sensor electrode on the glass substrate and the bonded silicon substrate causes it, because the humidity effect on the capacitance is the same when an oxide film (sputtered film, thickness 1000 Å) is deposited on the sensor electrode.
3. Even if the stem is changed or the extreme die-bond conditions to the stem are changed, the humidity effect is not changed significantly. Therefore, the stem is not main cause.
4. The variation of the humidity effect in the wafer shows a different trend to that of the process data in wafer processes such as deposition or etching. Furthermore, sensors in the same wafer lot have different humidity characteristics. Therefore, the silicon and glass processes do not cause it.
5. The humidity effect is small after cleaning the sensor in water, but that after heat treatment (523 K, 3 h) tends to be large. The effect depends on the default amount of water on the sensor surface. Accordingly, the adsorption of water molecules on the sensor surface mainly causes the humidity characteristic.
6. When the humidity characteristic is measured by changing the measuring frequency (100 kHz–1 MHz) of the impedance meter, a larger characteristic change is obtained as the measuring frequency is decreased. As the result, an electric element of high damping time constant seems to be involved.

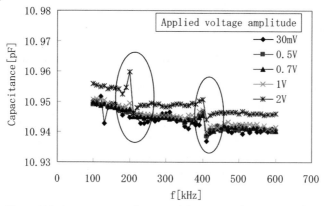

Figure 6.18 Capacitance as a function of frequency and amplitude of the applied voltage.

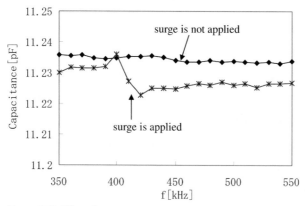

Figure 6.19 Effect of surge on capacitance. Measurements were made before and after the surge (pulse voltage) was applied.

6.8
Diaphragm Resonance

Figure 6.18 depicts the capacitance as a function of measurement frequency. Each plot shows the different applied voltages between the fixed and moveable electrodes. As can be seen, two remarkable changes in capacitance are observed (around $f = 200$ and $400\,$kHz). The resonance of the diaphragm seems to cause this capacitance change. Also, the thickness of the diaphragm changes the resonance frequency.

The maximum capacitance around $f = 200\,$kHz increases as the applied voltage increases. The resonant frequency is affected by electrostatic attraction. The amount of change increases with electrification by static electricity. The details of this are described in next section.

6.9
Surge (Static Electricity)

In a capacitive pressure sensor, the characteristics are significantly affected by electric damage because the gap is narrow, ~1 μm.

In the frequency dependence of the applied voltage (Figure 6.18) described above, the amount of capacitive change increases around a frequency of 400 kHz for the points where a remarkable capacitive change (around f = 200 and 400 kHz) was observed with an applied pulse voltage due to surge (static electricity, Figure 6.19).

It is necessary to take precautions not to change the capacitive characteristics by applying static electricity to the sensor chip through a factory worker.

6.10
Conclusion

A new silicon capacitive pressure sensor with center-clamped diaphragm structure has been developed. It was found by both FEM analysis and experiments that this sensor has better linearity. The non-linearity of this sensor is 0.7% FS when the ratio of changed to initial capacitance is 0.23, which is half that of a conventional flat-plate diaphragm sensor. In addition, a design for high mechanical reliability was proposed and fabricated. The evaluation of mechanical reliability showed excellent results. Furthermore, the humidity, resonance and surge characteristics were measured for high reliability. It is expected that the new diaphragm structure will be suitable for the realization of a high-performance, highly reliable and inexpensive sensor.

References

1 Y. S. Lee, K. D. Wise, A batch fabricated silicon pressure transducer with low temperature sensitivity, *IEEE Trans. Electron Devices* **ED-29** (1982) 48–50.

2 S. Shoji, T. Nisase, M. Esashi, T. Matsuo, Fabrication of an implantable capacitive type pressure sensor, *Proc. 4th Int. Conf. Solid-State Sensors and Actuators, 2–5 June 1987, Tokyo*, (1987) 305–308.

3 T. Kudoh, S. Shoji, M. Esashi, An integrated miniature capacitive pressure sensor, *Sens. Actuators A* **29** (1991) 185–193.

4 B. Pures, E. Peeters, W. Sansen, CAD tools in mechanical sensor design, *Sens. Actuators* **17** (1989) 423–429.

5 G. S. Sander, J. W. Knutti, J. D. Meidel, A monolithic capacitive pressure sensor with pulse-period output, *IEEE Trans. Electron Devices* **ED-27** (1980) 927–930.

6 K. Petersen, Silicon as a mechanical material, *Proc. IEEE* **70** (1982) 420–457.

7
Inertial Sensors

Osamu Torayashiki, Kenji Komaki, Sumitomo Precision Products Co., Ltd., Amagasaki, Japan

Abstract

MEMS Inertial Sensors are expanding its use from consumer and automotive applications into new applications which require higher performance. The reliability of the MEMS sensor is one of the major concerns in that situation. And it is important that how the reliability is built into the product. In this chapter, then, specific approaches to achieve the market requirements in reliability for our Capacitive-type Three-axis Accelerometer and Gyro sensors are discussed. Especially, some measures with regard to MEMS processes which we have done are presented.

Keywords

inertial sensors; gyro sensor; accelerometer; reliability

7.1
Introduction

Inertial sensors made by microelectromechanical systems (MEMS) technology have recently been widely utilized in several areas, such as automotive, aerospace,

Reliability of MEMS.
Edited by O. Tabata, T. Tsuchiya
Copyright © 2013 WILEY-VCH Verlag GmbH & Co. KGaA, Weinheim
ISBN: 978-3-527-33501-5

industrial and consumer applications. Typical examples of inertial sensors are gyro and accelerometer types. Originally these inertial sensors were intended only for aerospace applications and measurements as precision instruments. Subsequently, smaller and cheaper accelerometers with MEMS technology have been adopted for crash detection systems and applications are expanding to consumer products. Recently, smaller and cheaper gyro sensors have been used for image stabilization systems for camcorders and applications have spread to electrical stability programs (ESPs), navigation and roll-over detection in the automotive industry. While maintaining their small size and advantageous economic features, some MEMS sensors achieved have higher performance and reliability. Consequently, MEMS inertial sensors have already started to be used for aerospace and measurements, and especially their wider use in defense and space applications is also expected. One of the key aspects of the realization of such usage is reliability. However, there is no established standard procedure to validate MEMS reliability and its standardization has only just started. For instance, the US Army went ahead with a program to establish the guidelines for reliability testing of MEMS in military applications in 2004 [1]. On the other hand, acceptable reliability for MEMS products has already been demonstrated in automotive applications. Reliability should be accomplished with careful consideration of several aspects, such as design, testing and evaluation, based on the know-how and experience of individual developers and manufacturers.

In this chapter, we focus on MEMS gyro sensors and accelerometers from the point of view of reliability.

Standard procedures for reliability testing for automotive electronics is well established. One of the standards is the AEC-Q100 [2], in which automotive electronics is divided into five grades according to the operating temperature range:

- grade 0: −40 to +150 °C
- grade 1: −40 to +125 °C
- grade 2: −40 to +105 °C
- grade 3: −40 to +85 °C
- grade 4: 0 to +70 °C.

A gyro sensor for ESP requires grade 3 or 2, because the sensor is typically set in a console box. A gyro sensor assembled into an electronic control unit (ECU) requires grade 1, because the module is set in the engine room. For example, grade 1 parts should survive at −50 to +150 °C for 1000 cycles of temperature stress. Automotive sensors also require mechanical resistance, e.g. 0.5 ms of 1500 g shock and 50 g vibration up to 20 kHz. Even in these severe environments, a failure rate less than 1 in 10^6 is required for 15 years of life.

Environmental conditions and test procedures for airborne equipment are also provided in the standard RTCA/DO-160C [3]. In this standard, equipment is categorized by its maximum altitude. For instance, for a commercial aircraft operating at altitudes up to 35 000 ft, test conditions are in the range −55 to +85 °C and a low pressure test is also required.

Several approaches to developing a reliable sensor are necessary from the design stage through to the production stage. It seems reasonable to consider the following six categories:

- design
- materials
- manufacturing
- testing
- system
- evaluation.

In these approaches, the following failure modes should be considered in general:

- stiction
- wear
- electrostatic discharge
- contamination
- creep
- fracture/fatigue.

These failure modes are identified to establish guidelines for reliability testing for MEMS in military applications [1]. Because our MEMS inertial sensors do not have any contact point in the moveable portion in the standard operating mode, we excluded wear from our considerations. On the other hand, a key factor is to maintain the air pressure around the moveable parts in order to stabilize the resonant characteristics, because resonance is one of the features of MEMS. Therefore, air leakage was included among the typical failure modes and all six of the above failure modes should be considered in each approach. We divided the category of design into devices and processes. Table 7.1 shows the items considered for our MEMS inertial sensors.

Table 7.1 Matrix between items which should be considered and failure modes: the ticks show points which were considered for the MEMS inertial sensors.

Design	Stiction	Electrical discharge	Contamination	Creep	Fracture/ fatigue	Air leakage
Device	✓	✓	✓	✓	✓	✓
Process	✓		✓	✓	✓	✓
Material			✓	✓	✓	✓
Manufacturing			✓			✓
Test			✓			✓
Evaluation	✓	✓	✓	✓	✓	✓
System	✓	✓	✓	✓	✓	✓

All the six failure modes were then studied in the device design. To prevent stiction and electrical discharge, the use of bumps and/or insulation film at the contact point was studied in the design. The gap size at the moving portion was assessed by considering possible size. Minimization of the stress at the moving portion was also important from the points of view of creep and fracture/fatigue.

Stiction and contamination are closely related in process design. Particularly the wet process should be especially considered, because a moveable structure is already formed in the device. Creep and fracture/fatigue were also considered in process design. The materials should be selected with consideration of contamination, creep fracture/fatigue and air leakage. There are many possibilities for contamination of the device in the manufacturing stage. Although basically the cleaning process adopted is achieved using technologies and know-how which were developed in the IC industry, there are some specific cases for MEMS. Therefore, various ideas for preventing contamination in the manufacturing process were adopted.

Tests to inspect the failure caused by contamination and/or air leakage were installed in the line. Several evaluations were made in order to assure that no failure occurred in any failure mode during the life of the device. A diagnostic system was established assuming the occurrence of failure as the worst case. The actual approaches which we followed with our MEMS devices are described in the subsequent sections.

7.2
Capacitive-type Three-axis Accelerometer

This accelerometer was developed employing a bulk macromachining process with the anisotropic wet etching technique, which was a standard process to make a 3D structure with single-crystal silicon before the advent of deep reactive ion etching (DRIE). Even these days when we have DRIE technology for silicon, we cannot ignore this wet process because of its cost advantage. A countermeasure in the process to obtain suitable strength of the beam which was formed by anisotropic wet etching process is presented.

7.2.1
Principle of Acceleration Detection

Two kinds of accelerometer were developed. The first consists of four layers, a top glass with fixed electrode, a silicon moveable electrode with four beams, a glass seismic mass and stopper silicon [4], as shown in Figure 7.1. The Z sensitivity depends on both the stiffness of the beam and the mass. The X/Y sensitivity depends on the position of the center of gravity of the seismic mass in addition to the stiffness of the beam and mass. The glass seismic mass gives us freedom in the design to optimize the position of the center of gravity. Therefore, X, Y and Z of this sensor have equal sensitivity and resolution. The second type consists of

Figure 7.1 The three-axis capacitive-type accelerometer
(Sumitomo Precision Products), which consists of four layers:
top glass with fixed electrode, silicon moveable electrode with
four beams, glass seismic mass and stopper silicon.

Figure 7.2 Three-axis capacitive-type accelerometer, which
consists of three layers: top glass with fixed electrode, silicon
moveable electrode with four beams and with seismic mass
and stopper glass.

three layers, a top glass with fixed electrode, a silicon moveable electrode and a stopper glass, as shown in Figure 7.2. The beams and seismic mass are formed from silicon. This leads to a limitation in the design with respect to the position of the center of gravity, because there is a limit to the thickness of the silicon wafer due to wet anisotropic etching. However, this structure and process are simpler than in the previous case, and this device was developed as a high-volume and low-cost version of the four-layered type.

Figure 7.3 illustrates the principle of the accelerometer. Four electrodes are formed symmetrically on the fixed plate in order to separate acceleration vectors into X, Y and Z components. A flat common electrode with freedom of movement in the Z direction and twisting around the X and Y axes faces against the fixed electrode. The moveable electrode has a seismic mass below the electrode plane in order to generate rotating motion by X and/or Y acceleration.

Figure 7.4 shows the pattern of the fixed electrodes, C1–C5, and the outline of the moveable electrode with four beams. C1 and C3 are arranged symmetrically

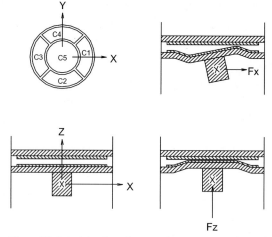

Figure 7.3 Principle of acceleration detection. Four electrodes are formed symmetrically on the fixed plate in order to separate acceleration vectors into X, Y and Z components.

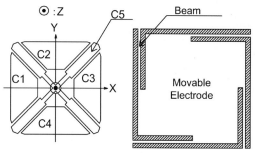

Figure 7.4 Schematics of the design of the fixed electrodes, C1–C5, and moveable electrode with four beams.

against the Y axis and C2 and C4 are arranged symmetrically against X axis. C5 has a symmetrical shape with respect to both the X and Y axes. The X (Y) component of the acceleration vector is obtained as the change in the difference between the C1 (C2) and C3 (C4) capacitance values and the Z component is detected directly as the change in the C5 capacitance value.

The change in value of the average gap distance between the fixed and moveable electrodes is linearly proportion to the applied acceleration. The capacitance value of the parallel plate is given by

$$C = \varepsilon S/d \tag{1}$$

where C is the capacitance, ε is the dielectric constant, S is the area of the electrode and d is the distance of the electrode. Hence the capacitance is inversely

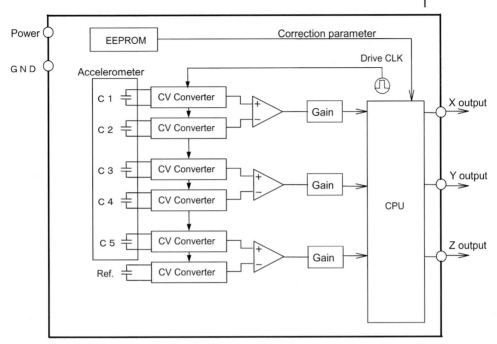

Figure 7.5 Block diagram of the capacitive-type three-axis accelerometer with correlation function.

proportional to the acceleration. Therefore, the simple difference in the capacitances includes systematic errors, such as non-linearity and cross-axis sensitivity. We have also developed a correlation method, which is a numerical method to obtain an accurate value of the acceleration from the output signal, the application of which requires higher accuracy [5]. This accelerometer with correlation function achieves less than 1% error with respect to non-linearity and cross-axis sensitivity.

Figure 7.5 shows a block diagram of this capacitive-type three-axis accelerometer with correlation function.

7.2.2
Structure, Materials and Process

Figure 7.6 shows the fabrication process flow of the three-layer capacitive type of three-axis accelerometer. The top surface of a double polished silicon wafer with (100) orientation is etched down to 5 μm with tetramethylammonium hydroxide (TMAH) with an SiO_2 mask, and pillars to maintain the capacitance gap are formed. The area which becomes the beams is etched down from the bottom side with TMAH and the seismic mass is also formed at the same time. Then the thickness of the etched portion is adjusted to 60 μm, as shown in Figure 7.6a and b. Then the silicon wafer is etched again from the top side with TMAH after pat-

(a)

(b)

(c)

(d)

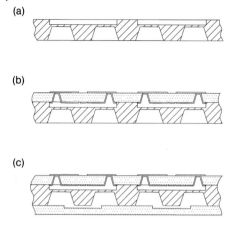

Figure 7.6 Fabrication process flow for the capacitive-type three-axis accelerometer.

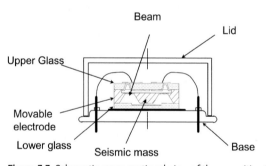

Figure 7.7 Schematic cross-sectional view of the capacitive-type three-axis accelerometer.

terning for the insulation layer and metal electrode. The silicon wafer is bonded anodically with both a top glass with electrodes and stopper glass, as shown in Figure 7.6c. The bonded wafer is cut out by dicing and assembled into a vacuum package (Figures 7.6d and 7.7).

The anisotropic wet etching of silicon with an alkaline solution is widely used for bulk MEMS, especially in pressure sensor manufacturing [6]. This technique applies a difference in the etching rate on each crystal orientation and a (111) surface appears at the side wall in the case of a (100) wafer, as shown in Figure 7.8. In the early stage of development we had a strength problem with the beam. Under moderate shock while processing, for example, the shock on hitting the

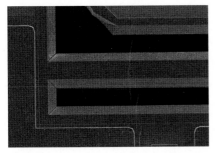

Figure 7.8 SEM image of the beam of the capacitive-type three-axis accelerometer formed by anisotropic wet etching.

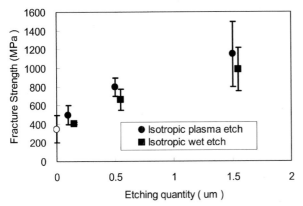

Figure 7.9 Experimental results for the improvement of the fracture strength by rounding by isotropic etching. The capacitive-type three-axis accelerometer device was used as the measurement specimen.

metal table with the wafer, some beams broke away. It was considered that the strength of the beam was weaker than expected. We made simple measurements to obtain the fracture strength of the beam as follows. We loaded force on to the moveable electrode using a push–pull gauge, which was an instrument used for measuring bonded wire strength, and recorded the maximum loading force when the beam broke. The average loading force was 3.5 g and some pieces broke only at 2 g. A level of 2 g corresponds to a fracture strength of 200 MPa, which was estimated from numerical analysis assuming the dimensions of the beam shape. We had expected the fracture strength to be around 600 MPa from previous work [7]. Because the cross-section of the beam has a trapezoidal shape, the top edge where the stress due to bending is mostly concentrated was the focus of consideration and two countermeasures were tried. The first was to round off the sharp corner of the edge which formed, by isotropic wet etching, and the second was to cons-trol the crystal defect at the surface of the silicon wafer. Figure 7.9 shows the experimental results of the effect of rounding by isotropic etching. Both isotropic etching with a CF_4 plasma and wet isotropic etching with a mixture of hydrofluoric acid, acetic acid and nitric acid significantly improved the fracture strength of the silicon beam. In order to make clear the effect of the crystal defect of silicon on the fracture strength, several experiments and observations were made. However, we could not find any consistent causal relationship between the existing crystal defect and the strength. Hence we established a reliable wafer process with isotropic etching treatment after anisotropic wet etching. A robust MEMS accelerometer was developed.

7.3
Inductive-type Gyro Sensor

An inductive type gyro sensor has been developed, targeting a yaw rate sensor for ESP. A life of 15 years with a wide operating range from –40 to 105 °C is required in addition to higher accuracy. The device developed is a kind of vibrating gyro sensor using the Coriolis effect of ring-shape oscillation [8]. Figure 7.10 shows a photograph of the sensing element without lid.

7.3.1
Principle of Operation

To use the Coriolis effect to detect angular rotation, the ring vibrator is forced to vibrate at its resonant frequency of $\cos 2\theta$ shown as mode 1 in the Figure 7.11. This vibration provides the velocity component of mass at the oscillating portion. When the whole ring is rotating, the Coriolis force (F_c) is generated in proportion to mass (m), velocity (v) and angular velocity (Ω) as follows.

Figure 7.10 Sensing element without lid of Silicon Sensing Systems inductive-type gyro sensor, which was developed in a project with collaboration with Sumitomo Precision Products and BAE Systems.

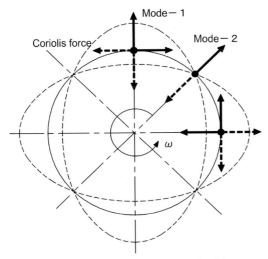

Figure 7.11 Schematic of the vibration mode of the ring.

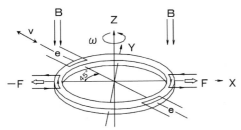

Figure 7.12 Schematic view of how to drive and detect the ring vibration of the inductive-type gyro sensor in practice.

$$F_c = 2mv\Omega \tag{2}$$

The Coriolis force generated by the angular rotation excites another vibration mode which is shown as mode 2 in Figure 7.11. If the velocity is kept constant, the amplitude of mode 2 is proportional to the applied angular velocity [8].

Figure 7.12 shows how to drive and detect the ring vibration in practice. Conducting electrodes, which form one turn coil, are provided on the ring at the oscillating position of mode 1 and a perpendicular magnetic field is applied through the conducting electrodes. Then the a.c. current applied to the conductors causes the ring to oscillate by Lorenz force. In order to detect the mode 2 vibration, other conducting electrodes are also provided at the oscillating portion of mode 2 which is the null point of mode 1. An induced voltage is created in the conducting electrode because this is a moving conductor in the magnetic field.

Figure 7.13 shows a block diagram of the operation of the sensor. In the primary loop the ring resonator excites with constant amplitude at its resonant frequency.

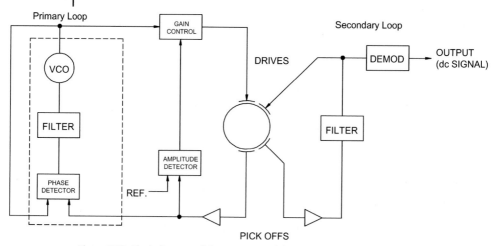

Figure 7.13 Block diagram of the operation of the ring structure vibrating gyro sensor.

In the secondary loop a rotation-induced secondary pick-off signal is detected and fed back to null the mode 2 vibration to zero. The output signal is obtained by demodulating the secondary drive signal to obtain a d.c. voltage, which is proportional to angular velocity [8].

7.3.2
Resonator Design

The ring resonator was designed with ring diameter 6 mm, ring width 120 µm, ring thickness 100 µm and resonant frequency 14 000 Hz.

Because the resonant motion is used on this sensor to obtain high sensitivity, it is desirable that the frequency split between modes 1 and 2 is close to zero. In practice, the allowance range of the frequency split depends on its quality factor (Q value). A high Q requires a small frequency split, whereas a low Q brings a wide allowance to the frequency split. However, a high Q value is suitable for high sensitivity. We adopted a high Q value with a vacuum package in order to develop a more accurate gyro sensor to meet the requirements for ESP application. Therefore, a ring resonator which has a symmetrical shape with high accuracy was required. That was accomplished by MEMS process technology, particularly photolithography having the capability of precise two-dimensional patterning and DRIE having the capability of anisotropic vertical etching. The frequency difference between the two modes, with a nominal resonant frequency of 14 000 Hz, was finally controlled to less than 1 Hz.

From the reliability point of view, it is clear that creep and fatigue causing a change in the mechanical properties of the resonator are one of the points to consider. Also, the reliability of the sealing, which should maintain the vacuum for life, stiction, contamination and fracture of the MEMS structure when shock is applied should also be considered carefully. In the next section, how we dealt with the reliability issues and what should be considered from the process point of view are discussed.

7.3.3
Process and Materials

The ring and legs are made from a single-crystal silicon wafer using photolithography and the DRIE technique. Figure 7.14 shows an aluminum alloy electrode patterned on a layer of silicon dioxide at the ring and legs. The silicon wafer is bonded with a pedestal glass wafer by anodic bonding. Magnetic parts are automatically assembled on a spacer glass wafer in parallel. Then the silicon wafer with pedestal glass and the spacer glass wafer are bonded. The magnet and cap are also assembled on the combined wafer. After dicing, each sensor element is picked up from the wafer to be assembled into a package. Figure 7.15 shows each component of the gyro sensor.

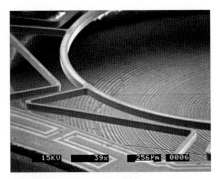

Figure 7.14 SEM image of the ring resonator and its legs.

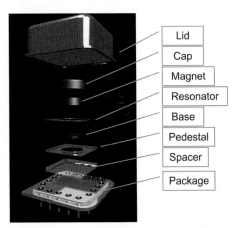

Figure 7.15 Schematic of the components of the inductive gyro sensor.

First, the stress of silicon should be considered. The typical amplitude of the ring is designed as 20 μm. The maximum stress of the moving portion is 65 MPa. Because we know that the fracture strength of silicon is more than 500 MPa, we can judge that the maximum stress level is small enough. Fatigue of silicon has been reported in several papers; however, it has also been reported that less than 50% of the fracture strength does not show any fatigue [9]. Therefore, we conclude that the fatigue of silicon is not an issue for this sensor.

Second, creep of the aluminum alloy film patterned on the ring should be considered. The aluminum electrode is formed on the silicon dioxide film using physical vapor deposition (PVD), photolithography and chemical etching, which are conventional processes in IC fabrication. Aluminum has a potential for creeping at not very high temperatures [10]. For instance, it is widely recognized as the stress migration of aluminum alloys in IC chips. In general, low resistance is one of the main requirements in IC technology where there are long wirings with narrow width. Soft metal is also preferred for easy wire bonding. However, it is also important for MEMS that the mechanical characteristics of the metal layer on the resonator are stable when applying periodic stress by oscillation.

In particular, we observed that the mechanical properties of aluminum alloy film that had been metallized by PVD changed on oscillation, resulting in a slight change in the resonant characteristics.

The metal properties obtained when using PVD are controllable by the process conditions, such as pressure, power, temperature and composition of the metal. We have optimized the metal properties and introduced the process for this gyro sensor. Thus when metal material is used in MEMS, due attention should be paid to creep and fatigue.

Third, vacuum should be considered. As mentioned earlier, the resonator of this gyro sensor is in a vacuum in order to obtain a resonator with a high Q value. This means that the ease of the oscillation is closely related to the pressure in the package. Therefore, it is necessary to guarantee that the inside pressure is maintained for life, because the resonator should always be oscillated with a constant amplitude by the driving force within a limited range.

Although it is easy to determine the pressure around resonator device by measuring the Q value, it is difficult to measure fine leakage from the package, as an acceptable leak rate, which comes from an acceptable change in the Q value during the life of the device, is too small to detect in the sealed package within the time frame allowed economically. The helium bomb method using a helium leak detector is a standard technique to check for leakage from a hermetically sealed package of an electrical device [11]. However, the detection sensitivity is two orders of magnitude less than the level we need. Therefore, this sealing process has been verified by accumulating a lot of process technology and know-how, such as a suitable design to obtain a stable sealing process, process conditions for a sufficiently wide process window and process control.

We used a resistive projection welding technique with a metal can package for this inductive-type gyro sensor, and this gyro sensor has proven to be sufficiently reliable for automotive applications.

Fourth, contamination should be considered. Because this device is not only capacitive but also inductive, there is no narrow gap in the design. The minimum gap size is 100 μm, which is between the upper/lower pole and ring surface. This is large enough compared with the particle size in the wafer process, but it is still a critical size for the particles existing in the can-type package which is produced by conventional machining and plating processes. Therefore, several special cleaning processes for the assembly parts and a further clean assembly process were developed. A highly reliable sensor was realized also by such contamination control in the manufacturing stage.

Finally, fracture at a high-g shock should be considered. The Bosch DRIE process is a relatively stable technique once the process conditions have been optimized. However, if the DRIE process is performed under critical conditions, several modes of unusual wall shape occur. Figure 7.16 shows SEM images of unusual wall shape.

The unusual shape generates some portion where stress is concentrated, hence fracture of the silicon structure is a potential failure. The problem is that the defect has various shapes that are unpredicted. Therefore, control of the etching process to avoid unusual shapes is required.

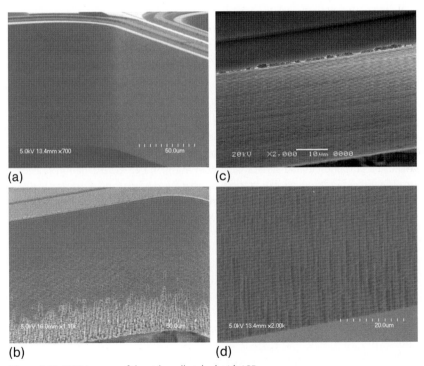

(a)

(b)

(c)

(d)

Figure 7.16 SEM images of the side wall etched with ICP etcher. (a) Smooth surface with no defect; (b) passivation breakdown; (c) "top hole" defect, which is a defect at the top edge; (d) striation.

Figure 7.16b shows a rough wall surface which is caused by breakdown of the passivation layer on the side wall. This phenomenon occurs when the ratio between deposition and etching is not optimized. Especially in the case of high aspect ratio and/or penetrating etching it has often been observed, because the ratio between deposition and etching changes as etching proceeds. Installation of parameter lumping, which is a method to maintain the deposition to etching ratio during the etching, is an effective measure.

Notching is one of the unusual side etch conditions at the bottom area [12]. Charging of the insulation film lying on the bottom of the etched silicon layer causes a notch and several measures to ensure suitable etching conditions have been proposed. These measures are very effective in avoiding notching, hence this is not an issue assuming that the etching time can be controlled within a safe range. Figure 7.16c shows striation and Figure 7.16d is an SEM image of the "top hole". These modes were observed in the case of a higher etching rate with higher r.f. power. These can also be avoided by the optimization of the process conditions, including the photoresist process. Establishment of a stable process which should have a sufficiently wide process window compared with the range of the total variation of the process is necessary.

7.4
Capacitive-type Gyro Sensor

A capacitive-type gyro sensor has been developed as a second-generation ring vibrating gyro, being motivated by cost and size requirements from the market [13]. Figure 7.17 shows the capacitive-type gyro sensor with ASIC assembled in a plastic package. Figure 7.18 shows the capacitive gyro sensor element, which consists of a glass–silicon–glass structure with silicon bulk micromachining.

The inductive detection and drive require precise three-dimensional assembly, because a perpendicular magnetic field B to the plane of the oscillation is necessary. On the other hand, a capacitive transducer for the drive and detection is easily implemented by electrodes arranged in the oscillation plane and fabricated by a wafer process using planar photolithography and DRIE. Also, the planar structure

Figure 7.17 Top view without lid of the Silicon Sensing Systems capacitive-type gyro sensor.

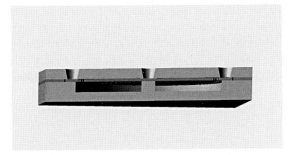

Figure 7.18 Schematic of the capacitive-type gyro sensor element which consists of glass–silicon–glass, developed in a project with collaboration with Sumitomo Precision Products and BAE Systems.

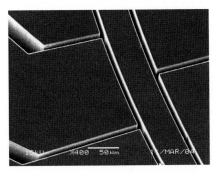

Figure 7.19 SEM image of the ring resonator and electrodes of the capacitive-type gyro.

is suitable for wafer level packaging (WLP) technology, which is an essential condition for a capacitive-type gyro sensor with narrow gaps from the point of view of preventing contamination. In another respect, this capacitive device has the potential for further shrinkage of the die size by implementing a finer design and process. Therefore, the concept of this capacitive-type gyro sensor has the potential for smaller size and lower cost for commercial and automotive sensors in the future.

The resonator of the capacitive gyro sensor is made only from a single-crystal silicon wafer: there is no other material on the resonator, in contrast to the inductive gyro sensor. The maximum stress of the vibration is also designed with a sufficient margin for silicon material. The resonator is also kept in a vacuum package for life. Therefore, there is no longer concern about the reliability as a resonator. Figure 7.19 shows an SEM image of the ring resonator and electrodes.

Because the capacitive drive and detection require a narrow gap structure, clogging at the gap should be considered more carefully than with the inductive-type gyro sensor. The gap size is designed as 10 mm and the amplitude of the ring is set at less than 5 μm. Therefore, contaminants larger than 5 mm should be excluded completely. In order to recognize such a clean process, first an all-dry process after forming the ring resonator and gap was introduced. Second, WLP to eliminate any contamination during the assembly process of the packaging was adopted. Thus a capacitive-type gyro sensor has been developed not only as low cost and with high performance, but also as a more reliable device.

7.5
System

Lastly, we mention some ideas for improving the reliability at the system level as examples. The situation that the system continues to work without any recognition of a failure when the sensor is out of order should be avoided. To install a set-up that detects such a failure is a realistic approach to achieving reliability at the system level. The build-in test (BIT), which is a self-diagnostic system, is also widely employed for the systems with high reliability. A self-diagnostic method using static electrical force was developed for a three-axis accelerometer. The moveable electrode is actually moved by applying static voltage to an additional electrode to simulate it, as if the sensor is subjected to acceleration. In this way, we can determine directly whether the sensor is working normally or not.

The vibration amplitude of the gyro sensor is controlled by the feedback of the primary pick-off signal. Therefore, by monitoring the frequency and drive current on the gyro system, we can also determine whether the ring is vibrating correctly with the expected Q value or not. Monitoring the Q value is the same as monitoring the pressure of the vacuum package. Hence a system-level approach using such features of each MEMS device is also important.

7.6
Conclusion

A MEMS inertial sensor should be considered in the six failure modes, stiction, electrostatic discharge, contamination, creep, fracture/fatigue and air leakage from the reliability point of view. Practical approaches to establishing highly reliable and high-volume MEMS sensors, such as the three-axis accelerometer, inductive-type gyro sensor and capacitive-type gyro sensor, were presented, focusing mainly on the MEMS process. The importance of the approach at the system level was also shown. Because especially the approaches at the system level using features of each MEMS device have a high possibility of improving the total reliability, these will be established in many systems. Regarding the test and evaluation method, activity aimed at its standardization has just started. Therefore, it is important to make great efforts to establish a standard for such a validation method.

7.7
Acknowledgment

The development of the gyro sensors, which has been done in collaboration with BAE Systems, owes much to helpful discussions with the team members.

References

1 R. Mason, L. Gintert, M. Rippen, D. Skelton, J. Zunino, I. Gutmanis, *Proc. SPIE* **6111** (2006) 61110K-1.

2 Automotive Electronics Council. *Stress Test Qualification for Integrated Circuits*, Component Technical Committee, AEC-Q100, 2003.

3 Radio Technical Commission for Aeronautics. *Environmental Conditions and Test Procedures for Airborne Equipment*, RTCA/DO-160C, 1989.

4 O. Torayashiki, A. Takahashi, R. Tokue, in *Technical Digest of the 14th Sensor Symposium, Japan*, 19–22, 1996.

5 Y. Araki, O. Torayashiki, A. Takahashi, R. Tokue, *Physical Sensors Technical Committee of IEEJ*, PS-96-15, 1996.

6 O. Tabata, R. Asahi, S. Sugiyama, in *Technical Digest of the 9th Sensor Symposium, Japan*, 15–18, 1990.

7 I. Igarashi, *Reports of Toyota Physical and Chemical Research Instituto*, **17** (1963) pp. 1.

8 I. Hopkin, presented at the DGON Symposium on Gyroscope Technology, Stuttgart, (1997).

9 T. Tsuchiya, A. Inoue, J. Sakata, M. Hashimoto, A. Yokoyama, M. Sugimoto, in *Technical Digest of the 16th Sensor Symposium, Japan*, 277–280, 1998.

10 D. S. Gardner, H. P. Longworth, P. A. Flinn, *J. Vac. Sci. Technol. A* **10** (1992) 1428–1441.

11 United States Department of Defense. *MIL-STD-833F, Method 1014.11*, a United States Defense Standard, 2004.

12 I. W. Rangelow, *J. Vac. Sci. Technol. A* **21** (2003) 1550–1562.

13 C. Fell, I. Hopkin, K. Townsend, I. Sturland, presented at the Symposium on Gyroscope Technology, Stuttgart, 1999.

8

High-accuracy, High-reliability MEMS Accelerometer

Matsushita Electric Works, Ltd., Osaka, Japan

Abstract

With the intention of improving the low-G acceleration sensor used in the automotive anti-lock brake system (ABS), a MEMS-based ultra-compact accelerometer has been developed by using an ICP etching process and devising a stress-relieving package structure and a circuit characteristics compensation method using digital trimming. The increased circuit gain to compensate for the decreased signal level derived from miniaturization negatively affected the output temperature characteristics and increased the output variation caused by thermal stress. However, the developed solution realizes a sensor chip 70% smaller than the previous model while delivering offset voltage–temperature characteristics (−40 to +85 °C) of ±2% full-scale (FS) or less and the characteristics change under an environmental test of ±3% FS or smaller.

Keywords

accelerometer; MEMS; reliability; accuracy; ABS

8.1
Introduction

In recent years, in the automotive industry, safety equipment, such as air bags and the anti-lock brake system (ABS), has becomes a standard feature. Therefore, the various kinds of sensor needed to control such safety equipments have also undergone rapid advances. Since the very small acceleration sensor which uses microelectromechanical systems (MEMS) technology is considered to respond to such needs, research and development in this area are highly active at present. Especially, an acceleration sensor for low G (~1.5 G) which is inexpensive and has high precision and high environmental resistance is strongly demanded. For several years, we have been producing a semiconductor piezoresistive acceleration sensor for the ABS system of four-wheel drive vehicles which has an acceleration range of $\pm9.8\,\mathrm{m\,s^{-2}}$ (~1 G) from the d.c. level. However, in order to respond to the rapidly advancing user needs, we have recently developed a new acceleration sensor, which has high precision and high reliability. The new acceleration sensor, fabricated using an inductively coupled plasma (ICP) etching process, realizes a size reduction of about 70% (size $1.5 \times 1.6\,\mathrm{mm}$) compared with the previous acceleration sensor chip [1] (size $2.6 \times 3.3\,\mathrm{mm}$), which is fabricated using anisotropic etching. In its development, the following problems which accompany the downsizing of a chip arose:

1. an increase in the offset voltage change due to the thermal stress
2. an increase in the offset voltage change due to the larger circuit gain
3. deterioration of the output temperature characteristic due to the sensor desensitization.

Therefore, to solve the above problems, we undertook (a) the improvement of the sensor chip and the package structure and (b) the development of a compensation (trimming) method for the temperature drift reduction, which is implemented in the readout integrated circuit (IC).

We developed the sensor chip and the package structure through which the thermal stress is difficult to spread. High environmental resistance was realized by reducing the offset voltage drift. Also, we developed a special application specific integrated circuit (ASIC), which includes digital trimming in offset and temperature characteristics compensation, and realized higher precision output characteristics and high environmental resistance by the adoption of a chopper amplifier. The high-precision and high-reliability acceleration sensor that was obtained with by the above-mentioned sensor chip and packaging and digital trimming IC is described in this chapter.

8.2
Acceleration Sensor Chip

8.2.1
Structure and Detection Principle of the Sensor Chip

The chip structure of the developed silicon piezoresistive accelerometer is shown schematically in Figure 8.1. This sensor chip is a cantilever-type accelerometer, in which a thick silicon mass is suspended by a thin silicon beam. Piezoresistive strain gauges are arranged on the beam surface using the diffusion process. Using high aspect ratio ICP-RIE for forming a cantilever, we attempted to reduce the size of the sensor chip. The silicon chip as sensing structure is covered by upper and lower glass substrates using anodic bonding in order to prevent cantilever failure on application of excessive acceleration and to protect against dust. The four strain gauges on the beam are connected in the form of a Wheatstone bridge, as shown in Figure 8.2, and when acceleration is applied, the cantilever mass moves up and down, and thus the beam bends upwards and downwards. The strain at the beam surface generated by the beam deflection causes a resistance change in the strain gauges, which results in the loss of balance in the Wheatstone

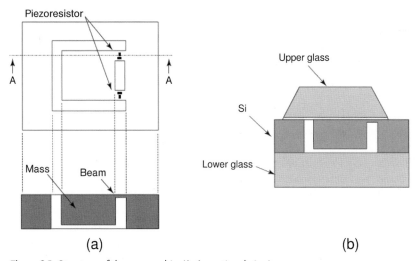

(a)   (b)

Figure 8.1 Structure of the sensor chip (A–A, sectional view).
(a) Si part (top, plane view; bottom, A–A sectional view).
(b) Sensor chip (A–A sectional view).

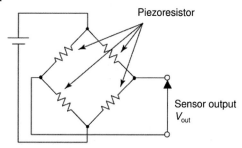

Figure 8.2 Equivalent circuit of the sensor chip.

bridge and a voltage output V_{out} appears. By amplifying the output using the detecting circuit, the sensor output voltage can be obtained.

8.2.2
The Issues for Controlling Disturbance Stress

In the semiconductor piezoresistive accelerometer, the warp caused by the beam bending against the application of acceleration changes the gauge resistances and, thus, the sensor output V_{out}. Therefore, the gauge resistance changes when the stress (hereafter called the disturbance stress) which occurred due to factors other than the acceleration on the sensor chip and V_{out} changes, which causes measurement errors and results in a decrease in the precision. Moreover, V_{out} may change during the long-term usage due to the drift of disturbance stress, and therefore high reliability cannot be achieved. The output caused by the disturbance stress appears as the sensor offset voltage that is added to the dynamically changing acceleration. Therefore, a new device was developed from the viewpoint of offset reduction, which contains (1) a structure in which the initial warp decreased significantly and (2) a structure free from disturbance stress, which means that the thermal stress does not transfer to beams in the sensor chip. The details of these two features are described in the following two sections.

8.2.3
Reduction of the Initial Warp of the Sensor Chip

There are various reasons for the initial warp of the sensor chip where the silicon structure was joined to the two glass structures on the upper and lower sides using anodic bonding. The major source of the offset voltage is the residual stress that causes the warp. The residual stress after the anodic bonding was caused by the difference in the coefficients of linear expansion of glass and silicon. Therefore, the initial warp of the sensor chip was reduced by the optimization of the shape of the bonding area using thermal stress analysis.

In this thermal stress analysis, three-dimensional stress analysis software was used. The stress caused by the anodic bonding is calculated by assuming that the

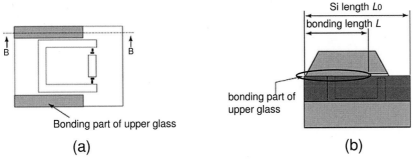

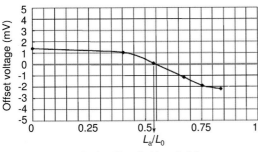

Figure 8.3 Structure of the sensor chip. (a) Si part plane view; (b) B–B, sectional view.

Figure 8.4 Calculated offset voltage generated from the thermal stress on the anodic bonding process. The offset voltage depends on the length of the bonded region.

silicon and glass structures are joined at 400 °C and cooled to room temperature. The stress at the strain gauges that is extracted from the analysis result is converted to the resistance change and thus the offset drift voltage is calculated. Since the strain gauges on the beams are located close to the area of bonding to the upper glass, the length L of the bonding part, as indicated in Figure 8.3, was selected as a parameter.

The optimum length where the offset voltage drift becomes zero was found as shown in Figure 8.4. When the upper glass was not bonded ($L = 0$), a positive offset voltage drift exists, which is caused by the thermal stress between the silicon and the lower glass. As the bonding length increased, the offset voltage drift decreased. At the optimum length L_a, the thermal stress from the lower and upper glass seemed to be balanced.

8.2.4
Disturbance Stress Reduction

Figure 8.5 shows a sectional view of the package structure. The sensor chip was bonded using soft silicone glue to the ceramic package that is mounted on the

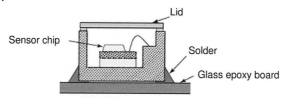

Figure 8.5 Packaging structure.

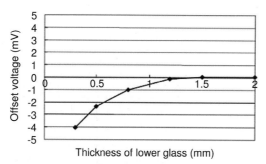

Thickness of lower glass (mm)

Figure 8.6 Calculated offset voltage in the operating temperature range, which is caused by the thermal stress of the assembled sensor. The thicker lower glass absorbs the package thermal stress.

glass epoxy circuit board using a solder as shown in Figure 8.5. When the ambient temperature of the mounted package is changed, the thermal stress propagates to the strain gauge on the sensor chip and causes the warp. If the thermal stress is too large, the plastic deformation at the bonding interface is induced, which results in output hysteresis, that is, temperature hysteresis of the offset voltage. In addition, when a large temperature cycle is applied, the plastic deformation is accumulated and a large offset drift may be observed.

Therefore, the package structure was optimized so as to reduce the output hysteresis by suppressing the transmission of the thermal stress of the package to the sensor chip and reducing the offset voltage drift.

To reduce the transmission of thermal stress, the following approach was adopted. First, we made the lower glass thick to absorb the thermal stress which occurs between the package and the sensor chip. This effect was simulated using three-dimensional thermal stress analysis, as described below. The stress distribution generated by the temperature change from 20 to 125 °C, which is identical with the actual environment, was converted to the offset voltage drift using the procedure described in the preceding subsection. In the analytical model, we computed the directly bonded sensor chip and the package and the die-bonding part (which has a stress absorbing effect) were not taken into consideration. The calculated result is shown in Figure 8.6. The stress-absorbing effect of increasing the thickness of the lower glass is remarkable and we found that the offset voltage

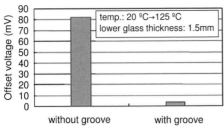

Figure 8.7 Calculated offset voltage drift of the mounted device with and without the stress-relieving groove.

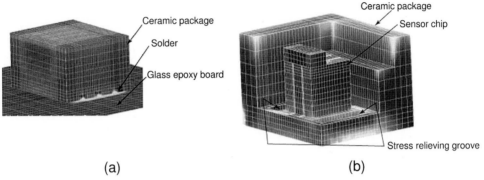

(a)   (b)

Figure 8.8 Mieses (principal) stress of package with the stress-relieving groove. (a) Stress on the outside surface of the package; (b) stress inside the package.

drift becomes almost zero at a glass thickness of 1.5 mm. As a result, we decided to use the 1.5-mm thick glass as the lower part of the sensor chip.

Second, to absorb the thermal stress which was generated by the thermal stress between the package and the glass epoxy board, which were bonded using a solder, we formed a stress-relieving groove on the bonding surface of the package base where the sensor chip is bonded. The stress reduction was verified by thermal stress analysis, as shown below. The calculated offset voltage drift of the mounted device with and without the stress-relieving groove and the stress distribution with the groove are shown in Figures 8.7 and 8.8, respectively. As shown in Figure 8.8, the stress distribution of the device with the stress relieving groove became low, especially near the sensor chip. Therefore, as shown in Figure 8.7, the offset voltage drift of the device with the stress-relieving groove decreased to 4% of that without the groove.

8.2.5
Experimental Results

The sensor chip and the package were fabricated and assembled based on the above design and the offset voltage change due to the thermal stress was mea-

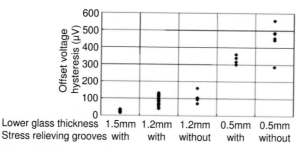

Figure 8.9 Measured offset voltage hysteresis as a function of the thickness of the lower glass and stress relieving grooves.

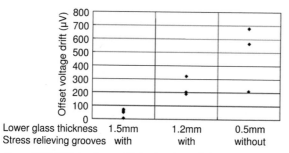

Figure 8.10 Offset voltage drift by the thermal shock test as a function of the thickness of the lower glass and stress-relieving grooves.

sured. The measurement results of an output hysteresis that is the offset voltage drift on exposure to the temperature cycle $(20 \rightarrow 85 \rightarrow 20 \rightarrow -40 \rightarrow 20\,°C)$ is shown in Figure 8.9. The output hysteresis was defined as the larger difference in offset voltage at the second 20°C and the third 20°C relative to the offset voltage at the first 20°C. Also, the measured offset voltage drift in the thermal shock test (100 to −40°C, in air) is shown in Figure 8.10. This confirmed the effect of the lower glass thickening and the groove on the package, and also the correlation between the output hysteresis and the offset voltage change.

8.3
Digital Trimming IC

8.3.1
Overview

In the semiconductor strain gauge type of acceleration sensor, the output voltage of the sensor chip itself is only several or several tens of millivolts per $9.8\,\mathrm{m\,s^{-2}}$ (~1 G) and the temperature coefficient of output also is as large as several tens of

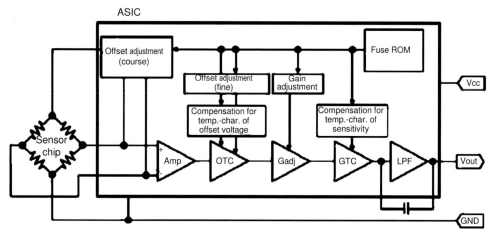

Figure 8.11 Block diagram of the readout circuit with digital trimming IC.

percent. On the other hand, according to the standard specification of the ABS acceleration sensor, the sensitivity is 1–2 V per $9.8\,\mathrm{m\,s^{-2}}$ (~1 G) and the accuracy is smaller than a few percent. Therefore, the electric circuit must have functions of higher gain, gain and offset adjustment and temperature compensation to the offset and gain. In the previously developed acceleration sensor, a laser trimmed thick-film resistor formed on a ceramic substrate was used in order to realize these adjusting functions. However, since the newly developed sensor chip is downsized to one-third of the previous sensor chip and the sensitivity decreases to about one-third, we need a three times higher gain of the amplifier to obtain the same sensitivity as with the previous acceleration sensor. In addition, in order to achieve high precision output characteristics and high gain at the same time, a high-precision regulating circuit that supplies a stable offset voltage and gain and a compensating circuit that cancels the temperature coefficient of the offset voltage and the sensitivity are required. Moreover, since the circuit has high gain, it tends to be influenced by the environmental stress. Therefore, higher reliability is needed with regard to the environmental resistance. In the development of the new acceleration sensor, not only the sensor chip but also an ASIC that has high-precision adjustment and high reliability were developed. The IC uses a digital trimming circuit. The internal block diagram of the IC is shown in Figure 8.11.

8.3.2
Issues for the Readout Circuit

As mentioned above, thick-film resistors on the ceramic substrate were used for trimming of the readout circuit adjustment. However, the resistors tend to drift in practical use because minute cracks (microcracks) formed in the thick-film resistor during the laser trimming grow during the heating and cooling cycles [2].

The growth of the microcracks causes drift of the offset voltage and circuit gain and deterioration of the temperature characteristic. Therefore, it is difficult to achieve high reliability of the acceleration sensor over the extended period. In addition, because the ASIC is formed on a silicon wafer in addition to the sensor chip, the offset voltage of the amplifier changes against the external stress, which is amplified and appears as the offset voltage drift.

8.3.3
Characteristics of the Developed Readout Circuit

The following are the three main features of the newly developed ASIC for readout circuits:

1. offset and gain trimming using a programmable read-only memory (PROM)
2. chopper amplifier of long-term stability on the offset voltage
3. quadratic compensation of the temperature offset and temperature coefficient of sensitivity for high-precision output characteristics.

8.3.3.1 PROM Gain Trimming

A fuse-type PROM was integrated in the ASIC in order to compensate for the characteristics of the sensor. The trimming data were calculated using the data obtained from the initial measurement to obtain the best circuit gain and the offset voltage. Although an electronically erasable and programmable read-only memory (EEPROM) can be used as a PROM for digital trimming, we adopted the fuse ROM, because the data are hardly rewritten or erased and the acceleration sensor is designed for automotive applications in which the sensor is used under severe environments. Since the adjustment data are held as digital data, the stability of offset and gain trimming is significantly improved in comparison with analog trimming such as the resistance value of the conventional thick-film resistance.

8.3.3.2 Chopper Amplifier

Using the chopper amplifier, high reliability was achieved, although the circuit gain of the sensor is about 500–1000, three times higher than that of the previous sensor.

8.3.3.3 Quadratic Compensation

The temperature coefficient of both offset and sensitivity were cancelled using the quadratic compensation function. The readout ASIC has a voltage source that creates a voltage V_t. The voltage source itself has a temperature coefficient. Therefore, the temperature coefficient of V_t is used for the temperature compensation of the sensor by adding the properly amplified V_t to the output voltage of the sensor chip. Then, the second voltage source V_{t2} which had a second-order temperature coefficient by multiplying V_t by itself was obtained. By adding V_{t2} to the output voltage of the sensor chip which was amplified similarly to V_t, temperature compensation of the second-order offset becomes possible and a high-precision output voltage, which is even higher than for the previous acceleration sensor, was achieved.

8.4
Experimental Results

8.4.1
Device Structure

The developed acceleration sensor for ABS is shown in Figure 8.12. The printed circuit board (Figure 8.13), on which the sensor chip and ASIC are mounted, is assembled in a plastic resin housing with a connector, which protects against dust and moisture from outside. The acceleration sensor is fixed to a metal bracket by clamping and is installed in the vehicle. The resin housing has input–output pins molded together, which have the role of the support and fixation of the printed circuit board and the electrical connections of the input–output signals.

Figure 8.12 ABS acceleration sensor.

Figure 8.13 Assembled circuit board.

8.4.2
Measurement Results

The experimentally fabricated acceleration sensor for ABS was subjected to a reliability test. The initial properties and the offset drift due to temperature are shown in Table 8.1 and Figure 8.14. As shown in Figure 8.14, an offset drift due to temperature of ±2% FS or less over a wide temperature range from 40 to 85 °C was achieved. The reliability test results and the offset drift by the thermal shock test are shown in Table 8.2 and Figure 8.15. As shown in Figure 8.15, an offset drift

Table 8.1 Initial properties of the fabricated acceleration sensor.

Item	Result
Acceleration range	±14.7 m s^{-2} (±1.5 G)
Operating temperature	−40 to 85 °C
Offset voltage	2.500 ± 0.060 V
Output voltage (+1.5 G)	4.000 ± 0.090 V
Output voltage (−1.5 G)	1.000 ± 0.090 V
Response frequency	0–20 Hz (−3 dB)

Table 8.2 Summary of the reliability test results.

Test item	Test conditions	P/F[a]
Low-temperature storage	−40 °C, 500 h	P
Low-temperature operation	−40 °C, 5 V, 168 h	P
High-temperature storage	125 °C, 500 h	P
High-temperature operation	85 °C, 5 V, 168 h	P
Heat shock	100 ↔ −40 °C/500 cycles	P
High-temperature and high-humidity operation	60 °C, 90% RH, 5 V, 300 h	P
Dew condensation	−5 ↔ 35 °C, 85% RH/5 cycles	P

a P, pass; F, failed. Evaluation criterion: offset drift of ±3% FS or less.

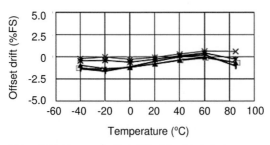

Figure 8.14 Measured offset drift of the fabricated sensors as a function of the operating temperature.

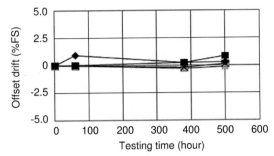

Figure 8.15 Reliability of the sensor. The offset drift by the thermal shock test.

by the thermal shock test of ±2% FS or less over a long testing time from 0 to 500 h was achieved.

As above, high resistance in the severe environment of automotive application was demonstrated.

8.5
Conclusion

A miniaturized and highly reliable piezoresistive acceleration sensor with digital trimming circuit has been developed. The shape of the upper glass cap for protecting the silicon device, especially the length of anodic bonding area, was optimized to minimize the offset voltage. In addition, a stress-relieving structure in a ceramic package and a readout ASIC with a digital trimming function and a chopper amplifier were developed. As a result, an offset voltage drift of ±3% FS or less in various reliability tests such as the thermal shock test was achieved.

An offset drift due to temperature of ±2% FS or less over a wide temperature range from 40 to 85 °C was achieved by adopting quadratic compensation functions.

In the future, we plan to expand variations of the product using these techniques and to spread out into other markets.

References

1 T. Ishida, K. Kataoka, H. Kami, H. Saito, S. Akai, K. Nohara, *Semiconductor Acceleration Sensor Device*, Matsushita Electric Works Technical Report No. 70, Matsushita Electric Works, Osaka, 49–54, 2000.

2 Y. Fukuoka, Trimming, in *The Hybrid Microelectronics Handbook*, Kogyo Chosakai Publishing, Tokyo, Ch. 8, Sect. 3, 463–467, 1989.

9
Reliability of MEMS Variable Optical Attenuator

Keiji Isamoto, Changho Chong, Santec Corporation, Aichi, Japan
Hiroshi Toshiyoshi, Institute of Industrial Science, University of Tokyo, Japan

Abstract

This chapter deals with an electrostatic tilt mirror, which was developed for fiber-optic variable attenuator, and discusses on its reliability issues such as mechanical strength of silicon suspensions, vibration tolerance, temperature dependence of electrostatic actuation, electrostatic drift, and stiction problems.

Keywords

variable optical attenuator; MEMS; electrostatic actuator; wavelength division multiplexing

9.1
Introduction

The variable optical attenuator (VOA) is an indispensable component in fiber-optic communication network systems and a large number of VOA components

Reliability of MEMS.
Edited by O. Tabata, T. Tsuchiya
Copyright © 2013 WILEY-VCH Verlag GmbH & Co. KGaA, Weinheim
ISBN: 978-3-527-33501-5

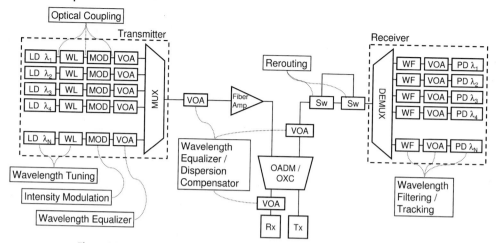

Figure 9.1 VOAs in optical fiber network.

are used particularly in the wavelength division multiplexing (WDM) system, where transmission light of multiple wavelengths (currently 8–16 channels) is individually controlled by the wavelength equalizer before or after wavelength multiplexing (MUX) and demultiplexing (DEMUX). VOAs are also used to tune the light intensity from laser diodes to fiber amplifiers and to photodetectors, as shown in Figure 9.1. The role of VOA is to maintain the quality of fiber-optic communication, i.e. to suppress the bit error rate by adjusting the intensity of traveling light in the appropriate range such that the optoelectronic devices function with their optimum performance.

The most conventional type of VOA consists of an optical component such as a prism or a mirror driven by an electromagnetic motor. Despite excellent optical performance, the large package volume does not meet the demands of the WDM system, in which multiple VOA components are accommodated in a limited space. Apart from the small size, conventional VOAs have not met customers' requirement such as low power consumption, low-voltage operation, high mechanical stability and low-cost productivity; microscale electromechanical systems (MEMS) approaches are expected to give a solution to these demands.

Figure 9.2 compares three different architectures of MEMS VOAs. The shutter insertion type [1–5] shown in Figure 9.2a can be easily integrated by surface- or bulk-micromachined actuators, but polarization-dependent loss (PDL) is generally larger than those of other types. Rotating or sliding a mirror in the coupled optical beam [6], as shown in Figure 9.2b, can also control the coupling efficiency, but it tends to occupy a relatively large area to hold the optical fibers on the substrate. Besides moveable mirrors, electromechanically actuated Fabry–Pérot interferometers can also be used to control attenuation. However, complete optical block-out (over 40 dB attenuation) is difficult to achieve by optical interference [7]. For these reasons, we employed an electrostatic torsion mirror coupled with two fibers through a collimator lens, as shown in Figure 9.2c. The advantages of this opto-

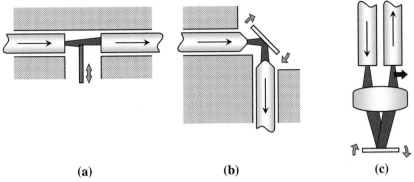

(a)   (b)   (c)

Figure 9.2 MEMS approaches to fiber-optic VOA.

mechanical design are small PDL, small package size and scalability to arrayed multi-channels for WDM applications [8].

9.2
MEMS VOA Design and Fabrication

Simplicity of mechanisms and fabrication was the first priority in designing the VOA mirror and we concluded that the most suitable design would be a parallel-plate electrostatic torsion mirror. Figure 9.3a illustrates schematically the VOA mirror structure. A circular mirror with electrostatic parallel plates is suspended by a pair of silicon torsion bars over the silicon substrate with a shaped through hole. The upper structure including the mirror and the suspension is made of silicon-on-insulator (SOI) 30 μm thick and the bottom part is the handle wafer. The right-hand side of the actuator plate has a shaped counter electrode (part of the substrate) with a 2 μm thick air gap. Applying voltage to the suspended structure generates electrostatic torque to tilt the mirror clockwise. The dimensions of the supporting hinges are designed such that the mirror would more likely tilt than move laterally. An analytical model for the suspension's elastic deformation is discussed in Section 9.4.2.

A large through hole was made in the substrate to keep the mirror from hitting the substrate when tilted at a large angle. The optical design requires a mechanical tilt angle of only 0.4° for 40 dB attenuation, provided that a pair of single-mode optical fibers (at a 125 μm pitch) is placed at the focal length of the collimator lens (1.8 mm focal length). As shown in Figure 9.3b, the maximum angle of the mirror is limited by the contact angle of the actuator, g/W_a, while the controllable angle range is limited by the electrostatic pull-in. Considering the operation margin, we set the maximum operation angle to be smaller than one-third of the contact angle.

Figure 9.4 illustrates the fabrication steps using two deep reactive-ion etching (DRIE) processes. In step 1, an SOI layer (30 μm thick silicon) on buried oxide (BOX, 2 μm thick silicon oxide) was first patterned into the shape of the scanner

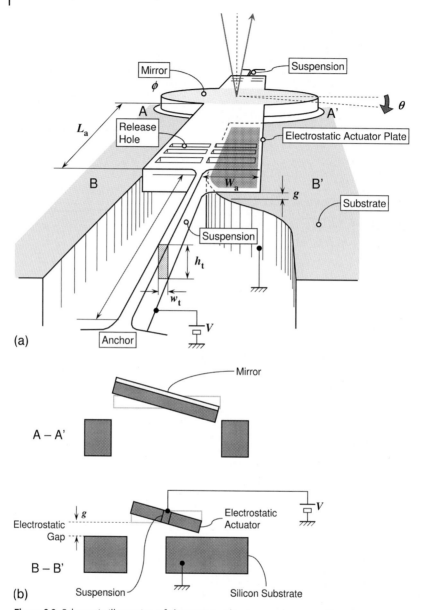

Figure 9.3 Schematic illustration of electrostatic tilt mirror and cross-section views.

by DRIE using a photoresist (Shipley S-1818 spun at 3000 rpm for 30 s). We used the DRIE machine manufactured by STS with a so-called "SOI option" process capability that minimizes the side-etching at the SOI–BOX interface (champagne glass effect) [9]. In step–2, the backside aluminum (100 nm thick) was etched into the shape of the chip. After protecting the front surface with another thick photo-

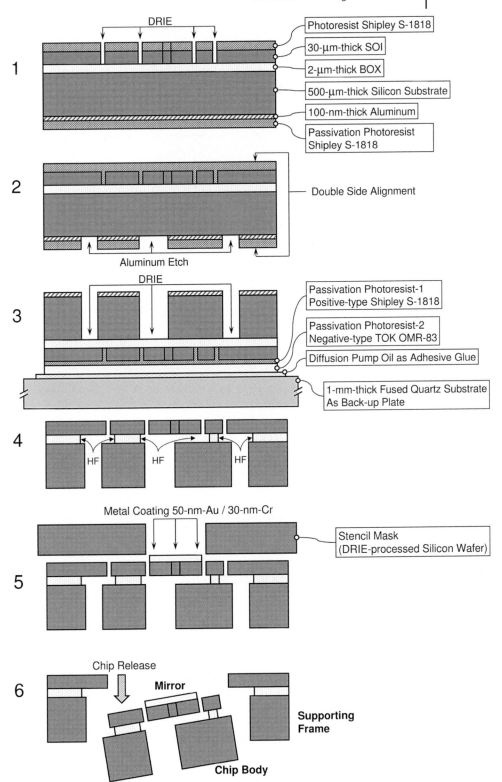

Figure 9.4 Simplified steps of silicon micromachining for MEMS VOA mirror.

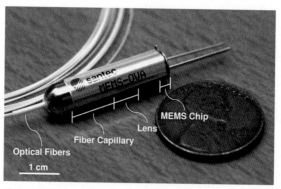

Figure 9.5 MEMS VOA package after optical assembly.

resist layer, in step 3 the wafer was flipped upside-down and bonded to a 1 mm thick quartz substrate for the subsequent through-wafer DRIE process. Diffusion-pump (DP) oil [10] was found to be a suitable material to fix the silicon wafer temporarily on the quartz wafer thanks to its adequate thermal conductance. The backside of the silicon wafer was etched to the BOX using a 100 nm thick aluminum mask. The wafer after the backside DRIE step was slid off the quartz wafer and washed with acetone to remove the DP oil and the passivation photoresist. After selectively removing the BOX with hydrofluoric acid in step 4, a 30 nm thick chromium and a 50 nm thick gold layer were deposited by vacuum evaporation through a stencil mask in step 5. We took special care to apply sacrificial release of the chip at a high processing yield; we will come back to this technical issue later. Finally, in step 6, the chips were snapped off the frame for device packaging. The total package size was as small as 5.6 mm in diameter and 23 mm in length, including an optical fiber capillary and a collimator lens, as shown in Figure 9.5. The VOA has been already commercially released in both single- and eight-channel packages.

Figure 9.6a shows a scanning electron microscope (SEM) image of the developed torsion mirror chip (2.4 × 2.4 mm in area and 0.5 mm in thickness) mounted on a TO-3 package stem. The center mirror was 0.6 mm in diameter and the actuator plates on the sides were each 700 μm long. The typical radius of curvature of the mirror was measured to be 2 m or longer after metal coating. Figure 9.6b shows a close-up view of the suspension (2 μm wide and 400 μm long), whose supporting roots were intentionally rounded at a radius of 50 μm for mechanical strength. Thanks to this shape, the suspensions were found to survive under mechanical shock upwards of 500 G (~5000 m s^{-2}). Tiny bars extending out from the torsion hinge were designed to prevent the SOI and BOX membranes from rupturing during the fabrication process. The parallel slits seen in the actuator plates were the release holes to let the HF acid to the underlying BOX layer for sacrificial release. Despite the delicate structure, the fabrication yield was nearly 100% for 250 chips in a 4-inch SOI wafer. Design parameters are listed in Table 9.1.

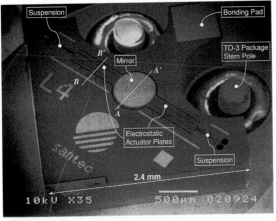

(a)

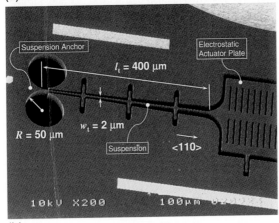

(b)

Figure 9.6 SEM image of MEMS VOA mirror and close-up view of the suspension.

Table 9.1 Design parameters of MEMS VOA mirror.

Parameter	Value	Notes
Mirror diameter, ϕ	600	
Actuator plate length, L_a	700	
Actuator plate width, W_a	80–90	To be tailored for various contact angles
Suspension length, l_t	150–400	To be tailored for various operation voltages
Suspension width, w_t	1.6	Measured
Structure height, h_t	30	SOI thickness
Electrostatic gap, g	2.0 ± 0.1	BOX thickness
Substrate thickness, h_{sub}	525	

9.3
Optomechanical Performance

Figure 9.7 plots a typical angle–voltage curve of the developed scanner that was characterized by the laser Doppler vibrometer. The actuator dimensions were designed to have the pull-in angle greater than the operation angle, 0.4°. After the pull-in around 5 V, the mirror was brought into contact with the counter electrode (handle substrate); electrical short-circuit was avoided due to the relatively high substrate electric resistivity (1–10 Ω cm). Low-voltage operation was also desirable to lower the risk of electrical short-circuit. After contact, the mirror was released at a voltage around 2.3 V.

Low-voltage operation was made possible partly because the suspensions were designed to be thin for a small elastic constant. However, the inertia mass was designed to be small such that the mechanical resonant frequency was as high as 1 kHz, as shown in Figure 9.8 (measured with 3 V d.c. with a 1 V a.c. signal). Thanks to this design, the response time was found to be in the range of a few milliseconds, as plotted in Figure 9.9. Due to the high quality factor (Q-factor ≈ 10) found in the frequency response in Figure 9.8, the mirror was found to have an overshoot for a step input voltage (settle time ~2 ms). A damped oscillation was also found at the falling edge in the mirror's response (settle time ~3 ms). The millisecond response is acceptable for fiber-optic communication systems.

The MEMS chip was assembled into the VOA package (shown in Figure 9.5) and the optical attenuation curve was examined as a function of applied d.c. voltage, as shown in Figure 9.10. The mirror position and the fiber position were aligned to have minimum insertion loss (~0.8 dB) by using the optical alignment machine equipped with a YAG laser welding tool. A maximum of 40 dB attenuation was obtained with a voltage lower than 5 V; this was useful to block completely the light for digital on–off control of signals. Attenuation repeatability was also inspected; the typical repeatability error was found to be less than 0.1 dB for an

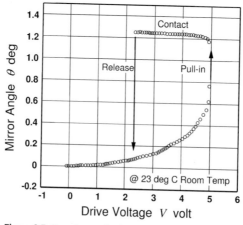

Figure 9.7 Experimentally determined mirror angle as a function of drive voltage.

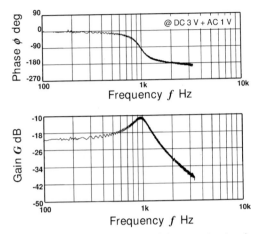

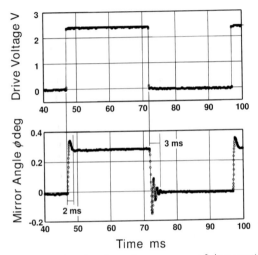

Figure 9.8 Frequency response of electrostatic mirror's angle.

Figure 9.9 Step-function voltage response of electrostatic mirror.

attenuation range lower than 20 dB. Detail VOA performances are summarized in Table 9.2.

9.4
Discussion on Reliability

9.4.1
Mechanical Strength

The first technical challenge in our project was to improve the mechanical strength of the torsion bars, such that the device would not break during the fabrication

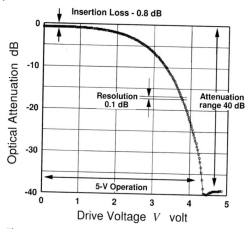

Figure 9.10 Optical attenuation as a function of voltage.

Table 9.2 Performance of MEMS VOA.

Parameter	Typical value	Notes
Drive voltage	5 V d.c.	
Insertion loss (IL)	0.8 dB	At 0 V
Attenuation range	40 dB	Blocking (>45 dB)
Resolution	0.1 dB	Open-loop control
Repeatability	0.1 dB	Open-loop control
Wavelength-dependent loss (WDL)	0.8 dB	At 20 dB attenuation
Polarization-dependent loss (PDL)	0.2 dB	At 20 dB attenuation
Polarization mode dispersion (PMD)	0.1 ps max.	
Response	<5 ms	
Optical power handling	300 mW max.	
Temperature control	Not required	Using a calibrated look-up table
Operating temperature	−5 to 70 °C	
Power consumption	10 mW max.	

process or optical assembly. After careful failure analysis, we frequently found the torsion bars to be broken near their supporting root at the anchors. The device was also found to be particularly weak when it was shaken in the horizontal direction. In our later mask designs, we employed a side-wall stopper that worked to limit the lateral motion of the actuator plate, as shown in Figure 9.11. The gap between the suspension and the wall was designed to be as small as 2 μm, which was close to the resolution limit of the silicon DRIE for the 30 μm thick structure.

The tail of the side-wall stopper was intentionally extended along the suspension root by a quarter turn (typical radius of curvature 50 μm) to protect the torsion bar

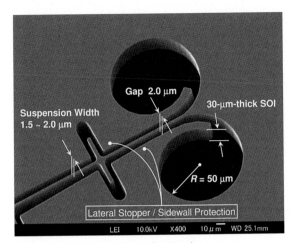

Figure 9.11 Supporting hinge of torsion bar.

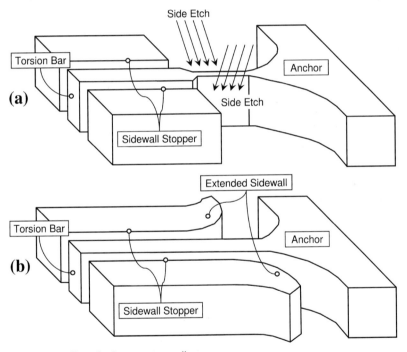

Figure 9.12 Effect of side-protection wall.

during the DRIE process. Figure 9.12 illustrates schematically how the torsion bar was etched by the DRIE process. When the protection side-wall was short, the suspension root was directly exposed to the etching plasma, as shown in Figure 9.12a, and the suspension root became thinner due to the side-etching effect of DRIE. This resulted in stress concentration at the neck, when the torsion bars were excessively bent. In contrast, we newly integrated the extended side-wall protection as shown in Figure 9.12b; thanks to the protection function, no suspension necking was observed and the scanner was found to survive after external vibration upwards of 500 G given by the shock test bench.

9.4.2
Low-voltage Operation and Vibration Tolerance

The scanner was designed to operate in the low-voltage range of ≤5 V. Low-voltage operation was made possible by increasing the actuator's torque and by reducing the torsion bar's rigidity. Due to the limited chip size, the actuator area could not be expanded for larger torque. Therefore, we mainly used thin torsion bars for low-voltage design. We found, however, that low-voltage operation was possible at the risk of degraded device stability to external vibration. As plotted in Figure 9.13, one of the first VOA prototypes (suspension width 2 μm, length 200 μm) was found to be too compliant to the external shock of 25 G; the attenuation level (preset to 10 dB) was disturbed by 7 dB, which was not acceptable for an optical communication system.

The mechanism of electrostatic force/torque coupled with the external vibration is shown in Figure 9.14. The actuator plates were expected to remain at the initial gap g at drive voltage V, giving the tilt motion θ only. When external vibration in the Z direction was applied, however, the actuator's center of gravity was displaced towards the substrate, decreasing the effective electrostatic gap to $g - \Delta g$. Hence the electrostatic torque became larger and it gave an additional tilt angle $\Delta\theta$ to the mirror. A solution to decouple the external vibration was to use torsion bars that were rigid in the bending motion for mechanical stiffness and that were compliant in the torsional motion for low-voltage operation.

The electromechanical model of the scanner was understood by considering the suspension's twist motion and the bending motion independently, as illustrated in Figure 9.15. From an elastic point of view, superposition of two types of

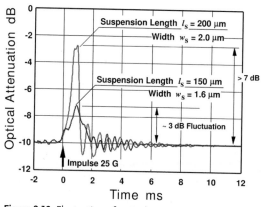

Figure 9.13 Fluctuation of optical attenuation under mechanical vibration.

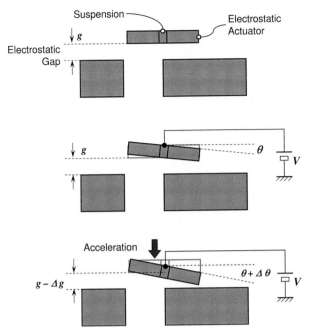

Figure 9.14 Electromechanical coupling with vibration.

$$k_\theta = 2 \times \frac{G \cdot h_t \cdot w_t^3}{3\,l_t} \left[1 - \frac{192}{\pi^5} \cdot \frac{w_t}{h_t} \cdot \tanh\left(\frac{\pi\,h_t}{2\,w_t} \right) \right]$$

$$k_z = 2 \times \frac{E_s \cdot w_t \cdot h_t^3}{l_t^3}$$

Figure 9.15 Suspension rigidities in rotation and bending.

mechanical deformation is acceptable in such a small displacement range. First, the mechanical restoring torque around the suspension, T_{elec}, and the electrostatic attraction force normal to the substrate, F_{elec}, are modeled by the following equations:

$$T_{elec}(\theta, z, V) = 2\eta \int_0^{W_a} \frac{1}{2} \varepsilon_0 \frac{L_a x}{\left(\dfrac{g-z}{\sin\theta} - x\right)\theta} V^2 dx \tag{1}$$

and

$$F_{elec}(\theta, z, V) = 2\eta \int_0^{W_a} \frac{1}{2} \varepsilon_0 \frac{L_a \cos\theta}{\left(\dfrac{g-z}{\sin\theta} - x\right)\theta} V^2 dx \tag{2}$$

where ε_0 is the dielectric constant of vacuum, $8.85 \times 10^{-12}\,\mathrm{F\,m^{-1}}$ and L_a and W_a are the length and width of the actuator plate (single side), respectively [8]. The x-axis of the coordinates is taken along the actuator's width with its origin ($x = 0$) being located on the rotational axis. The coefficient η is the relative effective area of the actuator plate that has vacant slits as etching holes; each release hole was designed to be $5 \times 60\,\mu\mathrm{m}$ located in every $15 \times 90\,\mu\mathrm{m}$ unit cell, so the coefficient η is calculated to be $[(15 \times 90) - (5 \times 60)\,\mu\mathrm{m}]/(15 \times 90\,\mu\mathrm{m}) = 0.77$.

On the other hand, the mechanical restoring torque, T_{mech}, and attraction force, F_{mech}, are written as

$$T_{mech}(\theta) = 2\frac{G h_t w_t^3}{3 l_t}\left[1 - \frac{192}{\pi^5}\frac{w_t}{h_t}\tanh\left(\frac{\pi h_t}{2 w_t}\right)\right]\theta \tag{3}$$

and

$$F_{mech}(z) = 2\frac{12 E I}{l_t^3} z, \quad I = \frac{w_t h_t^3}{12} \tag{4}$$

respectively, where G is the modulus of rigidity and E is Young's modulus of silicon. One would find values of θ and z under the electromechanical equilibrium condition by numerically solving the following simultaneous equations:

$$\begin{cases} T_{elec}(\theta, z, V) = T_{mech}(\theta) \\ F_{elec}(\theta, z, V) = T_{mech}(z) \end{cases} \tag{5}$$

The solid curves in Figure 9.16 shows theoretically calculated angle–voltage curves by using the dimension parameters listed in Table 9.1. To investigate the air gap-dependent variation, we used 1.9, 2.0 and 2.1 µm for the air gap g. The analytical model was found to agree with the experimental results (Figure 9.7) fairly well.

From Eqs. (3) and (4), one would obtain the contrast of rigidity, R, as

$$R = \frac{F_{mech}}{T_{mech}} \approx \frac{E}{G}\left(\frac{h_t}{w_t l_t}\right)^2 \tag{6}$$

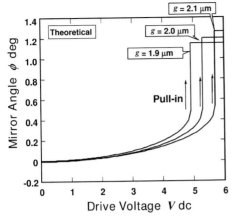

Figure 9.16 Analytical simulation of mirror angle as a function of voltage with different initial gap values.

The suspension height h_t is fixed to the SOI thickness and we have the suspension width w_t and length l_t as parameters to tailor the contrast of rigidities. Remembering that we would design high rigidity in the bending direction while seeking low-voltage operation, we would increase the ratio R by using thinner and shorter suspensions. Compared with the first design parameters, $w_t = 2\,\mu m$ and $l_t = 200\,\mu m$ in Figure 9.14, the improved suspensions with $w_t = 1.6\,\mu m$ and $l_t = 150\,\mu m$ had a smaller fluctuation of 3 dB, as compared in Figure 9.13. In our latest model, fluctuation has been suppressed to only 0.05 dB thanks to the newly employed air-damping mechanism.

Equation (6) implies another approach to improvement by using a larger value of the Young's modulus E and a smaller value of the shear modulus of rigidity G. Single-crystalline silicon has interesting elastic characteristics in favor of this requirement. As illustrated schematically in Figure 9.17, the Young's modulus along the silicon' s crystallographic axis <110>

direction (~160 GPa) is known to be larger than that along the <110>

direction (~130 GPa). At the same time, the shear modulus of rigidity around the <110>

axis (~62 GPa) is smaller than that around the <110>

axis (~79 GPa). Therefore, aligning the torsion bar in the <110>

direction is beneficial to increase the rigidity contrast R for low-voltage and shock-resistive electrostatic operation.

9.4.3
Temperature Dependence

We tested a number of VOA chips on a TO-3 package at different temperatures to investigate the temperature-dependent loss (TDL) variation. Interpreting the attenuator behavior in terms of the mirror angle, we discovered that the mirror at

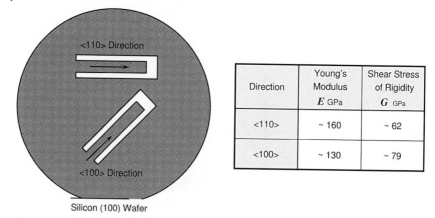

Direction	Young's Modulus E GPa	Shear Stress of Rigidity G GPa
<110>	~ 160	~ 62
<100>	~ 130	~ 79

Silicon (100) Wafer

Figure 9.17 Orientation-dependent elastic constants of silicon.

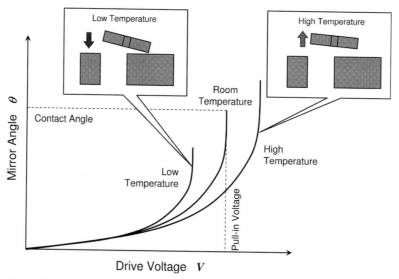

Figure 9.18 Temperature dependence of mirror angle–voltage curves.

low temperatures tilted more at a given voltage but with a smaller contact angle. In contrast, the mirror tilted less at higher temperature but with a larger contact angle. The phenomena are observed as shown in Figure 9.18.

A theory to explain the scanner's behavior was thermal unimorph deformation. In these devices, we used blanket metallization all over the released SOI layer by depositing a 30-nm chromium layer and a 50-nm gold layer by vacuum evaporation. The metal layers deposited at an elevated temperature occupied an expanded volume due to the thermal expansion coefficients. At room temperature after the vacuum deposition, the composite beam of silicon and metals changed

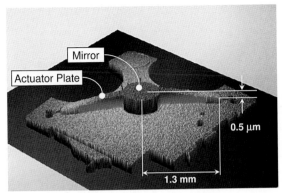

Figure 9.19 Optical interference microscope image showing mirror's buckling up.

its shape to a different radius of curvature, which changed the mean electrostatic gap g.

The first contribution to the thermoelastic deformation was found to be the built-in stress of the SOI layer. Confocal optical microscope observation revealed that the released mirror surface before metallization already had a buckling-up deformation, as shown in Figure 9.19. The total span including the torsion hinges was 2.6 mm and the peak-to-valley deformation of the mirror was 0.5 μm at room temperature. The deformation was attributed to the compressive stress of the SOI layer that had been thinned down by the chemical mechanical polishing (CMP) processes.

The thermal deformation after metallization can be understood by the electro-thermal model depicted in Figure 9.20. The SOI layer bonded on the handling wafer remains flat but with a built-in compressive stress due to the CMP process. Once the underlying BOX layer and substrate are removed, the SOI layer released the compressive stress by extending its length, i.e. by buckling up from the substrate. Covering the SOI surface with evaporated metals may partially compensate the stress. However, the effect would not be large enough to flatten the structure. Here, the temperature causes a change in the balance: at higher temperature, the metal layers expand more than the silicon does and the mirror disk bends downwards. The mirror occupies the largest portion in the suspended structures and the rest of the structure (including the actuator plates and the suspensions) are forced to follow the mirror's deformation to maintain a continuous surface curvature. Consequently, the actuator parts bend up from the substrate, giving additional space to the initial gap. Hence the mirror at an elevated temperature exhibited a contact angle which was larger than that at room temperature and a smaller angle at a lower temperature.

Based on the above assumptions, we speculated that the area (or the amount) of the deposited metal would have a significant effect on the thermal behavior of the scanner. We prepared VOA chips of identical design but with three different ways of metal coating: (a) a control sample with no metallization, (b) a sample

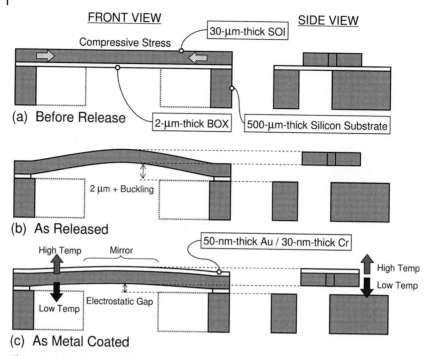

Figure 9.20 Mechanism of electromechanical coupling with thermal deformation.

with blanket-deposited metal; and (c) a sample metallized through a stencil mask on to the mirror part only. The chips were tested at various temperatures from −10 to 70 °C. Using the pull-in voltages was found to show quantitatively the temperature dependence, as plotted in Figure 9.21.

As expected, the sample without metallization exhibited the minimum dependence on temperature. The scanner with blanket metallization, on the other hand, was found to show the largest change. The scanner with a confined metallization area had an intermediate temperature dependence. From this observation, we concluded that the TDL would be minimized by confining the metal area to the mirror surface only. Figure 9.22 shows the stability of attenuation over wide temperature range from 0 to 70 °C. Thanks to the improved design, the 0 dB attenuation (intrinsic insertion loss, 0 V drive) had a variation of only ±0.15 dB. Even in the biased state (−20 dB attenuation), the variation was as small as 0.5 dB with open-loop control. In our latest VOA models, we also used metal deposition on the backside of the mirror to bring a balance to the thermomechanical stress in the mirror disk.

9.4.4
Electrostatic Drift

Most bottleneck problems in general MEMS device development are related to device packaging. MEMS devices working normally at the wafer or chip level

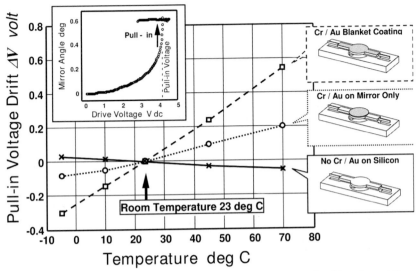

Figure 9.21 Temperature dependence of electrostatic mirror in terms of pull-in voltage drift.

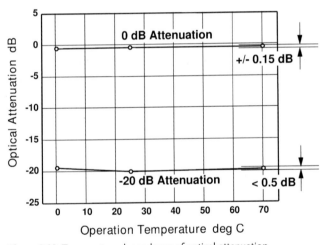

Figure 9.22 Temperature dependence of optical attenuation.

sometimes show unexpected behavior after they have been bonded or glued in a package. The source of the change can be found in the mechanical stress due to the bonding glue, which usually reduces its volume when it solidifies. Solder or silver paste is widely used to hold dies firmly but they are also known to show high post-process stress. We tested several methods of die fixation and finally used an epoxy glue of electrically insulating type due to the low stress.

The optical performance of the VOA exhibited long-term drift (time constant of a few minutes) and short-term fluctuations (in the range of a few seconds) even

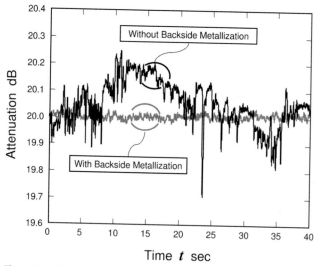

Figure 9.23 Fluctuation of VOA attenuation due to electrostatic drift.

though a constant voltage was applied to the MEMS chip; a typical time-dependent attenuation curve is shown in Figure 9.23. Such unstable behavior was found to occur randomly in sampled chips and the rate of incidence varied wafer by wafer. Even successfully passed VOAs were sometimes found to show unstable characteristics after a few days.

After carefully investigating the methods of die bonding and of wire bonding, we presumed that the source of fluctuation lies in the unstable electrical level on the chip substrate, in particular at the interface of the bottom surface and the glue. Figure 9.24a illustrates the electrical connection to the first prototype VOA chip. The SOI mirror was accessed via a metal contact pad, to which the gold wire was bonded. The substrate surface was also wire bonded in the same manner, where part of the SOI and BOX layers were removed. The backside of the substrate was bonded to the metal package stem; electrical ground connection was not made on the backside but on the top surface with wire bonding.

An equivalent circuit of the VOA can be understood by using Figure 9.24b. The electrical resistivity of the substrate was intentionally chosen to be high such that the electromechanical contact of the actuator would not be welded by electrical short-circuit. The leak resistance though the BOX was found to be as high as several tens of MΩ. The leak current through the distributed substrate resistance charges the stray capacitance between the substrate's back surface and the package stem. Due to the electrical resistance of the glue, the charges were trapped in the stray capacitance at node A to cause fluctuations of electrical potential, resulting in an unstable electrostatic force, i.e. the mirror angle.

The stability of the electrostatic operation was improved simply by reducing the electrical resistance of the substrate back. A thin metal layer (5 nm thick chromium

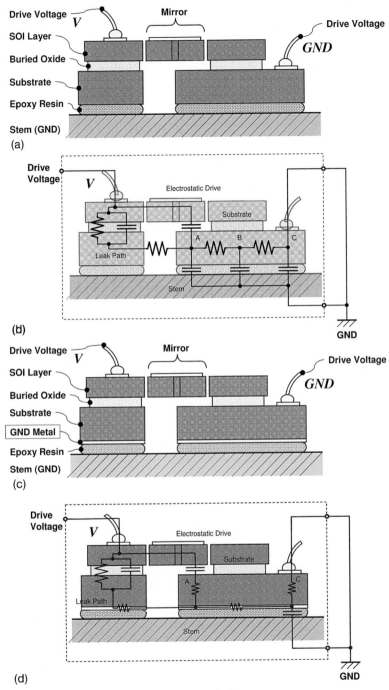

Figure 9.24 Electrical interconnection model (a) before improvement and (b) its equivalent circuit. (c) Model after improvement with backside metal layer and (d) its equivalent circuit.

and 50 nm thick gold) was vacuum evaporated on to the surface to flatten the electrical ground level of the substrate. The substrate was also connected by using the gold wire on the top surface, because it was found to give more reliable results than using electrically conducting glue. The VOA attenuation level after modification is shown in the same plot, Figure 9.23. Fluctuation was suppressed to ±0.05 dB, which corresponded to a mirror angle of ±0.0003°. As mentioned in the previous section, the backside metal deposition was also found to be effective in stabilizing the temperature-dependent loss.

9.4.5
Process Residue

Silicon wafers processed by the DRIE commonly suffer from dust particles generated by the micromasking effect. The DRIE process is a repetition of silicon etching and side-wall protection; particularly when the process condition is tuned for a vertical side-wall, the etching process is likely to fail to remove the protection material completely from the etched bottom, leaving micromasks that result in particles or silicon needles of sub-micron diameter. Etched silicon surfaces processed under such conditions can be identified by their dark-brown color. The presence of such microparticles was found to be harmful for the operation of the VOA scanner due to the risk of a particle being trapped in the electrostatic gap and limiting the tilt angle.

For post-process cleaning, we first used a mixture of 100 mL of sulfuric acid (96% H_2SO_4) and 100 mL of hydrogen peroxide (30% H_2O_2) to remove organic contamination; this is also known as SPM or piranha cleaning. A silicon wafer with MEMS device was dipped into the SPM beaker for 10 min and the hot-plate was set to keep the liquid temperature at 115 °C. After rinsing for 10 min in running deionized (DI) water, the wafer was dipped into hydrofluoric acid (10%) for 1 min to partially etch the BOX and to wash out the particles in a manner similar to the lift-off process. At this moment, the MEMS scanner was not released but remained pinned down. Further cleaning was carried out by using a combination of ammonia solution (29% NH_4OH) and hydrogen peroxide (30% H_2O_2) at 60 °C for 10 min; this is also known as APM cleaning, commonly used in CMOS processes. The ammonia cleaning was followed by rinsing for 10 min in running DI water. Detailed process conditions are summarized in Table 9.3.

9.4.6
Vapor HF Release Process

Wafers cleaned by the recipe in the previous section can be sacrificially released in wet hydrofluoric acid (HF). VOA mirror models of low-voltage operation were, however, difficult to release without causing process stiction. Most pieces could be salvaged by mechanically poking the mirror edge by using a probe micromanipulator, but the process throughput was extremely low.

Table 9.3 Post-process wafer cleaning.

No.	Process	Chemicals	Temperature (°C)	Duration (min)
1	SPM cleaning	30% hydrogen peroxide 100 mL96% sulfuric acid 100 mL	115	10
2	Rinse	DI water	Room	10
3	Partial etch	10% hydrofluoric acid	Room	1
4	Rinse	DI water	Room Temp.	5
5	APM cleaning	30% hydrogen peroxide 100 mL29% ammonia 100 mL	60	10
6	Rinse	DI water	Room	10

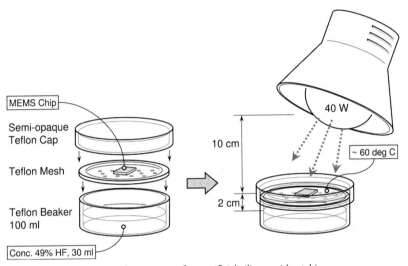

Figure 9.25 Utensil set-up for HF vapor for sacrificial silicon oxide etching.

As an alternative method, we used HF vapor to remove the silicon oxide selectively in the gas phase [11]. The HF vapor process was done without using any complicated vacuum system but with a set of simple plastic vessels placed in a draft chamber, as shown in Figure 9.25. A small amount of concentrated HF (50%, 30 mL) was put in a Teflon beaker (100 mL) and a MEMS chip (2 × 2 cm) was set on a plastic mesh and suspended over the liquid surface. A Teflon cap was placed on the beaker to retain the vapor of HF; the silicon oxide exposed to the HF vapor was selectively removed by the following reaction:

$$SiO_2 + 4HF \rightarrow SiF_4 \text{ (gas)} + H_2O \text{ (gas)}$$

Note that both SiF_4 and H_2O are volatile at room temperature. To assist the removal of water, we used the heat from a light bulb (40 W) located 10 cm from the Teflon cap. When the light bulb was set further than 10 cm away and the reaction temperature was lower, the moisture was not removed effectively and the etching result was similar to that of wet etching. When the light bulb was closer than 10 cm, on the other hand, the reaction temperature became too high and the moisture was instantly removed and the etching rate became low due to the lack of the catalytic effect of water. This observation implied the presence of an optimal temperature that could be controlled by an alternative method such as using a heater in the reaction chamber.

Figure 9.26 plots the side etch length as a function of reaction time. The typical etching rate (~0.12 µm min^{-1} for the first 1 h) with HF vapor was found to be equivalent to that with wet BHF (buffered hydrofluoric acid, 17%). The best result was obtained by using a stress-free SOI wafer (10 µm thick SOI); a maximum 5 mm long cantilever was successfully released without stiction, as shown in Figure 9.27 [11].

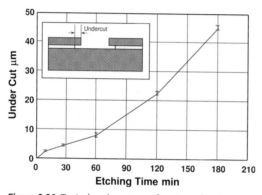

Figure 9.26 Typical undercut as a function of etching time.

10-microns-thick SOI beam, maximum length 5 millimeters

Figure 9.27 Silicon cantilevers after HF vapor sacrificial etching.

9.4.7
Mechanical Solution to In-use Stiction

Due to the structure of the electrostatic actuation mechanism, the actuator plate shown in Figure 9.3b makes physical contact with the substrate when an excess voltage is applied. The first prototype model was designed to touch the substrate with the entire actuator edge; this occasionally resulted in permanent contact of the actuator plate so that the mirror plate did not return to the flat position even when the drive voltage was reset to zero; the problem is usually referred to as in-use stiction.

In-use stiction was avoided by reducing the contact area of the actuator, as used in the Texas Instruments DMD [12]. We employed a similar point-contact structure as shown in Figure 9.28; the radius of the nail was designed to be 5 μm on the photomasks. The nail structure was found to lower the risk of permanent stiction to almost 1/5 that of the first prototype model.

9.4.8
Chemical Solution to In-use Stiction

Reducing the surface tension force or the surface free energy is essential to lower the risk of in-use stiction [13]. We used a self-assembled monolayer (SAM) from hexamethyldisilazane (HMDS) precursor, which is usually known as an adhesion promoter (primer) to positive-type photoresist. Figure 9.29 shows a modified fabrication process to release the VOA mirror. In step 1, the sacrificial silicon oxide is selectively removed in the HF vapor (described in Section 9.4.6), after which the silicon atom on the surface are terminated by hydrogen atoms (Si–H bond). The wafer is baked under a light bulb to evaporate the chemical products. In step 2, the wafer is exposed to ozone gas (O_3) at room temperature; the hydrogen atoms

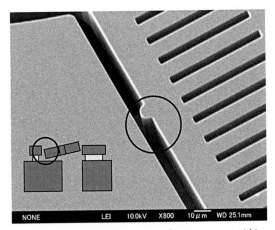

Figure 9.28 Silicon point-contact nail structure to avoid in-use stiction.

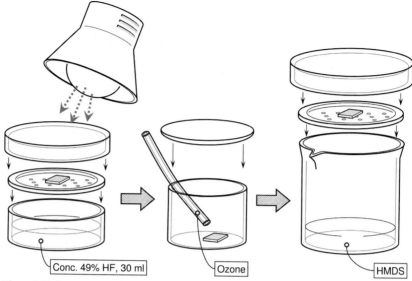

Conc. 49% HF, 30 ml | Ozone | HMDS

Figure 9.29 Process steps for anti-stiction surface coating.

on the surface are replaced with the hydroxide base (Si–OH bond). Finally, in step 3, the hydroxide-terminated silicon surface is exposed to the vapor of HMDS and the silicon surface is then terminated with a silane-coupled methyl molecule by the following reaction:

$$2SiOH + NH(SiMe_3)_2 \rightarrow 2SiOSiMe_3 + NH_3$$

where Me represents a methyl group ($-CH_3$).

VOA chips after each processing step were sampled to check the surface stiction. We measured the mirror angle as a function of applied voltage and used the release voltage as an index to measure the surface stiction quantitatively; in this section, we defined the release voltage as the voltage at which the mirror was released from the counter electrode after pull-in contact, as indicated by the lateral bar in Figure 9.30a. The release voltage can be interpreted as the stiction force such that the stronger the surface stiction is, the lower the release voltage becomes. Permanent stiction could be defined by the mirror in contact with the counter electrode at zero voltage. Throughout this investigation, the substrate part of the chip was electrically grounded and the suspended mirror structure was positively biased by drive voltage.

Figure 9.30b compares the release voltages of identical VOA chips that were sequentially processed in HF vapor, ozone and HMDS. First, a MEMS chip that had been exposed in ambient air for more than 1 week (labeled "as delivered") showed a relatively high release voltage of 1.8 V. The chip was exposed to the HF

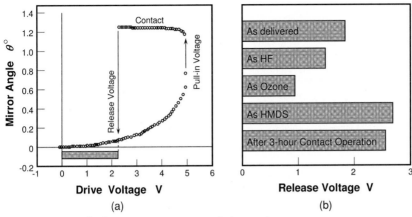

Figure 9.30 Strength of in-use stiction in terms of release voltage.

vapor for 5 min at room temperature, emulating an as-released chip, was found to have a lower release voltage, 1.5 V; this is a sign of increased adhesion force between the actuator and the counter electrode substrate. After processing the chip with ozone gas for 5 min, the release voltage became smaller. Finally, the chip was brought into the HMDS vapor for 5 min, after which the release voltage exhibited a drastic increase to 2.6 V or higher; this implies that the contact surfaces would have less tendency to undergo stiction. The effect of the SAM coating was found to last for more than 3 h of the contact experiment, which was equivalent to 54 000 cycles of contact.

The VOA scanner is usually operated within the mirror angle before pull-in without causing physical contact. The surface modification technique presented here can be used to assure that the operation is maintained when the actuator is accidentally pulled-in with excess drive voltage or external vibration. The reliability of the SAM coating is under investigation, in particular concerning the tempera-ture effect; the technique presented here is at the stage of laboratory-level research and we have not yet used it with commercially available products.

9.4.9
Hermetic Seal

Hermetic sealing of the MEMS device was found be indispensable to assure the reliability of actuation. The main purpose of the hermetic seal is to exclude mois-ture. In the VOA package, we use YAG laser welding of the stainless-steel case in a dry ambient atmosphere (nitrogen). After aligning the optical fiber capillary, the hermetic seal was quality tested in a vacuum chamber. A poor hermetic seal resulted in a loss of pressure in the package, changing the air damping effect on the MEMS scanner; therefore, observation of the mirror's dynamic performance could be used to rule out such poor packages.

References

1 V. Aksyuk, B. Barber, C. R. Giles, R. Ruel, L. Stulz, D. Bishop, Low insertion loss packaged and fibre connectorised MEMS reflective optical switch, *Electron. Lett.* **34** (1998) 1413–1414.

2 W. Noell, P.-A. Clerc, L. Dellmann, B. Guldimann, H.-P. Herzig, O. Manzardo, C. R. Marxer, K. J. Weible, R. Dandliker, N. De Rooij, Application of SOI-based optical MEMS, *IEEE J. Sel. Top. Quantum Electron.* **8** (2002) 148–154.

3 C. Marxer, B. de Jong, N. de Rooij, Comparison of MEMS variable optical attenuator designs, in *Proceedings of the 2002 IEEE/LEOS International Conference on Optical MEMS*, 55–56, 2002.

4 C.-H. Ji, Y. Yee, J. Choi, J.-U. Bu, Electromagnetic variable optical attenuator, in *Proceedings of the 2002 IEEE/LEOS International Conference on Optical MEMS*, 49–50, 2002.

5 M. G. Kim, J. H. Lee, A discrete positioning microactuator: linearity modeling and VOA application, in *Proceedings of the 13th International Conference on Solid-State Sensors, Actuators and Microsystems (Transducers 05)*, 984–987, 2005.

6 H. Cai, X. M. Zhang, C. Lu, A. Q. Liu, MEMS variable optical attenuator with linear attenuation using normal fibers, in *Proceedings of the 13th International Conference on Solid-State Sensors, Actuators and Microsystems (Transducers 05)*, 1171–1174, 2005.

7 K. W. Goossen, Microelectromechanical etalon modulator designs with wide angular tolerance for free-space optical links, *IEEE Photon. Tech. Lett.* **18** (2006) 959–961.

8 K. Isamoto, K. Kato, A. Morosawa, C. Chong, H. Fujita, H. Toshiyoshi, A 5-V operated MEMS variable optical attenuator by SOI bulk micromachining, *IEEE J. Sel. Top. Quantum Electron.* **10** (2004) 570–578.

9 Surface Technology Systems, Imperial Park, Newport, UK [http://www.stsystems.com/].

10 *Diffusion Pump Oil HIVAC-F-5 (500 mL)*, Shin-Etsu Chemical Co., Ltd., Tokyo, Japan.

11 Y. Fukuta, H. Fujita, H. Toshiyoshi, Vapor hydrofluoric acid sacrificial release technique for micro electro mechanical systems using labware, *Jpn. J. Appl. Phys.* **42** (*Part 1*) (2003) 3690–3694.

12 S. A. Henck, Lubrication of digital micromirror devices, *Tribol. Lett.* **3** (1997) 239–247.

13 R. Maboudian, R. T. Howe, Critical review: adhesion in surface micromechanical structures, *J. Vac. Sci. Technol. B* **15** (1997) 1–20.

10
Eco Scan MEMS Resonant Mirror

Yuzuru Ueda, Akira Yamazaki, Nippon Signal Co., Ltd., Saitama, Japan
Akira Yamazaki, Nippon Signal Co., Ltd., Tokyo, Japan

Abstract

Making use of the MEMS technology, we developed the MEMS resonant mirror "ECO SCAN", which is compact in size, light in weight and excels in quietness. The device has achieved low power consumption and large amplitude by making use of electromagnetic drive method and resonance phenomena. Moreover, it is also extremely durable thanks to its one-piece formed construction of single crystal silicon. Further explanations of the characters and potential applications to the area of optical goods are to be described in this paper.

Keywords

MEMS mirror; resonant mirror; Electromagnetic force; Lorentz force

Reliability of MEMS.
Edited by O. Tabata, T. Tsuchiya
Copyright © 2013 WILEY-VCH Verlag GmbH & Co. KGaA, Weinheim
ISBN: 978-3-527-33501-5

10.1
Introduction

Micro electromechanical systems (MEMS) resonant mirror devices have the potential to miniaturize optical systems, such as sensors, displays and printing devices. Since they can integrate the driving actuator into the mirror structure, a small, low power consumption and inexpensive device can be realized. A variety of driving methods, such as electrostatic, electromagnetic, piezoelectric and thermal types, has been developed for driving MEMS mirrors. Among them, electrostatic parallel plate actuators are widely adopted because of their simple fabrication. For example, the digital micromirror device (DMD) [1], which has been developed and put to practical use by Texas Instruments in the USA, drives a number of small mirrors placed on an integrated circuit chip using an electrostatic actuator, because the mirror is small enough and the structures are sufficiently simple to be suitable for arrayed devices.

However, it is difficult for electrostatic actuation to be applied with a large-sized mirror and a large rotation angle, because the electrostatic method requires the driving electrode to be placed adjacent to the mirror to generate a large electrostatic force. When we need large rotation angles, it is necessary to keep the distance between the mirror and the fixed electrode as long as possible to obtain a large rotation, which reduces the actuation force significantly. The electrostatic force is inversely proportional to the square of the distance between the mirror and the electrode, so the electrostatic method is insufficient to obtain a large amplitude. Electromagnetic force can generate a larger force than electrostatic force in relatively large (>100 µm) mirror devices and it does not require a fixed electrode, which shows that electromagnetic actuation is suitable for large rotation mirrors. A comparison between the methods of electromagnetic force and electrostatic force is summarized in Table 10.1.

To meet the demand for a device to be replaced with polygon mirrors and galvanometer scanners, we adopted electromagnetic actuation and developed a MEMS mirror [2, 3] in 1994, with which both the driving force and a large amplitude can be reliably obtained. Since then, we have been working on the practical application of these mirror devices. In this chapter, we introduce the Eco Scan MEMS resonant mirror in the respect of the operating principle, fabrication and basic char-

Table 10.1 Comparison between the methods of electromagnetic force and electrostatic force.

	Electromagnetic force	*Electrostatic force*
Principles of operation	Lorentz force	Electrostatic force generated by electrodes facing each other
Merits	Low-voltage driving Large amplitude obtainable Easy amplitude control	Suitable for downsizing (minimization)
Demerits	Magnet/yoke need to be implemented.Difficult to downsize due to the layout of the magnet and yoke	Large amplitude not obtainable High-voltage driving Difficult to control the amplitude The mirror size can not be large (less than 1 mm)
Practical accomplishments	Eco Scan	So far applied for acceleration sensors, gyro sensors, switches of optical communications, DMD, etc.

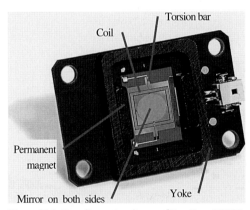

Figure 10.1 One-dimensional type of Eco Scan MEMS resonant mirror.

acteristics. We then describe the reliability of the Eco Scan, which is the most important property of mirror devices, especially the effect of damage in the fabrication process and fatigue life designs. Finally, we report some application systems using the Eco Scan.

10.2
Features

We have developed two types of the electromagnetic-type resonant mirror devices. Figure 10.1 shows the one-dimensional type of the Eco Scan MEMS resonant mirror [4] and Figure 10.2 shows the two-dimensional type. The Eco Scan is a device for changing the light path of the incoming light on to the mirror. The features are as follows:

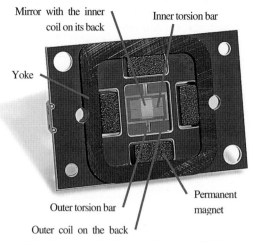

Mirror with the inner coil on its back

Inner torsion bar

Yoke

Permanent magnet

Outer torsion bar

Outer coil on the back

Figure 10.2 Two-dimensional type of Eco Scan MEMS resonant mirror.

1. large amplitude with low power consumption achieved by making use of the electromagnetic drive method and resonance vibration;
2. high-speed operation, extreme quietness and low power consumption achieved by reduction in size and weight;
3. optimum durability owing to the single-piece construction and preassembled nature of silicon microfabrication and the excellent mechanical properties of single-crystal silicon;
4. high productivity in mass production due to utilization of MEMS technology.

10.3
Principles of Operation

The Eco Scan MEMS resonant mirror uses the electromagnetic drive method. On a single-crystal silicon substrate, a mirror, torsion bars and coils are formed and permanent magnets are placed on the periphery of those parts. In the magnetic field of the permanent magnets, when a current is applied to the coil, Lorentz force is generated on the coil and thereby the mirror tilts to the position where the rotational torque of the coil is balanced by the restoring forces of the torsion bars. As the Lorentz force is proportional to the current, by changing the intensity of the current the tilting angle of the mirror can be adjusted, that is, the amplitude of light scanning can be controlled freely. Moreover, by applying alternating current matching the resonant frequency, a large amplitude can be obtained with low power consumption. One-dimensional scans can be performed with one pair of torsion bars whereas two pairs of torsion bars permit two-dimensional scans.

10.3.1
One-dimensional Type

Figure 10.3 shows the principle of operation of the one-dimensional type of Eco Scan. When a current i is applied to the coil with a magnetic field of magnetic flux density B applied in a direction perpendicular to the torsion bar, a rotational torque is generated by a Lorentz force F, thereby permitting the mirror to tilt to the position where the restoring forces of the torsion bars are balanced by the rotational torque. By changing the intensity of the current i, the tilting angle of the mirror can be adjusted, hence the amplitude of light scanning can be controlled freely as desired.

10.3.2
Two-dimensional Type

Figure 10.4 shows the principle of operation of the two-dimensional type of Eco Scan. When currents i_A and i_B are applied to the coils with a magnetic field of magnetic flux densities B_A and B_B in directions perpendicular to the torsion bars

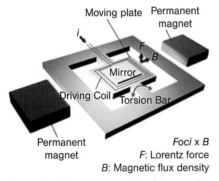

Figure 10.3 Operating principle of the one-dimensional type of mirror device.

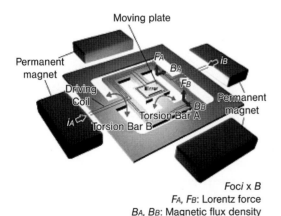

Figure 10.4 Operating principle of the two-dimensional type of mirror device.

A and B, respectively, rotational torques are generated by Lorentz forces F_A and F_B, respectively, thereby permitting the mirror to tilt to the position where the restoring forces of the torsion bars are balanced by the corresponding rotational torques. By changing the intensities of currents i_A and i_B, the tilting angle of the mirror can be adjusted, hence the amplitude of light scanning can be controlled freely as desired in the directions of two axes together or independently.

10.4
Manufacturing Process

10.4.1
One-dimensional Type

Figure 10.5 shows the manufacturing process of the one-dimensional type Eco Scan MEMS resonance mirror shown in Figure 10.1. An SOI (silicon on insulator) wafer is provided, which is thermally oxidized on both sides. The SOI wafer consists of three layers, namely a silicon active layer, a buried oxide layer of thermally oxidized film and a silicon support substrate. First, aluminum film is sputtered and then the first layer coil is formed by photolithography and etching (Figure 10.5a). Over the first layer coil, an interlayer insulation film is formed and then, in the same way as used for the first layer coil formation, the second layer coil is formed (Figure 10.5b). Then over the coil a passivation film is formed for corrosion protection (Figure 10.5c). In the next step, the silicon base is processed. To provide the torsion bar, the coil and the mirror, the unwanted parts of the thermally oxidized film, the active layer and the middle layer are removed (Figure 10.5d). Then, from the rear side of the SOI wafer, the unwanted parts of the thermally oxidized film, the support base and the buried oxide layer are removed and lastly both sides of the mirror are gold-plated (Figure 10.5e). The silicon wafer processed as shown in Figure 10.5 is cut to each chip by dicing and is assembled on the base together with the magnet and yoke.

10.4.2
Two-dimensional Type

Figure 10.6 shows the manufacturing process of the two-dimensional type Eco Scan MEMS resonance mirror shown in Figure 10.2. As in the case of the one-dimensional type, an SOI wafer is provided, which is thermally oxidized on both sides. First, aluminum film is sputtered and then the first layer coil is formed by photolithography and etching (Figure 10.6a). Over the first layer coil, an interlayer insulation film is formed and then, in the same way as used for the first layer coil formation, the second layer coil is formed (Figure 10.6b). Then another interlayer insulation film is formed over the second layer coil and the contacts, with which the coil wirings are connected, and a passivation film are formed for corrosion protection (Figure 10.6c). In the next step, the silicon base is processed. To provide

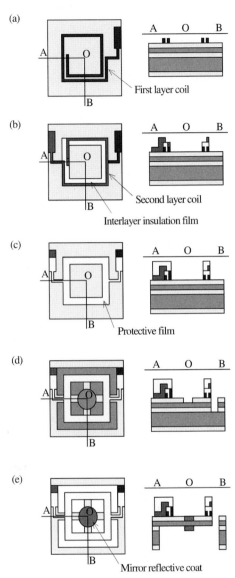

(a) First layer coil

(b) Second layer coil

Interlayer insulation film

(c) Protective film

(d)

(e) Mirror reflective coat

Figure 10.5 Manufacturing process of the one-dimensional type of mirror.

the torsion bars, the coil and the mirror, the unwanted parts of the thermally oxidized film, the active layer and the middle layer are removed (Figure 10.6d). Then, from the rear side of the SOI wafer, the unwanted parts of the thermally oxidized film, the support base and the buried oxide layer are removed and lastly both sides of the mirror are gold-plated (Figure 10.6e). The silicon wafer processed as shown in Figure 10.6 is cut to each chip by dicing and is assembled on the base together with the magnet and yoke.

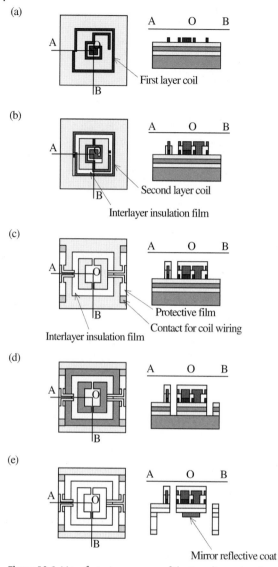

(a)

First layer coil

(b)

Second layer coil

Interlayer insulation film

(c)

Protective film

Contact for coil wiring

Interlayer insulation film

(d)

(e)

Mirror reflective coat

Figure 10.6 Manufacturing process of the two-dimensional type of mirror.

10.5
Operating Characteristics

As examples, the characteristics of the one-dimensional type (Figure 10.1) and the two-dimensional type (Figure 10.2) of Eco Scan MEMS resonant mirrors are given in Tables 10.2 and 10.3, respectively.

Table 10.2 Specification of the one-dimensional type of mirror.

Item		Specifications
Shape and dimensions	Mirror size width (torsion bar side) × length (other side)	5 × 6 mm
	Dimensions	24 × 37 × 5.6 mm
	Mass	9 g
Basic performance	Resonance frequency	540 Hz ± 5%
	Amplitude at resonance frequency (optical angle)	Less than ±34°
	Frequency–amplitude characteristics (sine wave driving)	See Figure 10.7
	Current–amplitude characteristics with amplitude at resonance frequency (sine wave driving)	See Figure 10.8
Optical characteristics	Material of mirror	Au
	Mirror reflectance	More than 85% ($\lambda = 670$ nm)
Electrical properties	Value of resistance	220 Ω± 10%
Lifetime	Number of scans: more than 1×10^9 times We have a practical accomplishment of 2-year consecutive driving test, which is equivalent to scanning 3×10^{10} times	

Table 10.3 Specification of the two-dimensional type of mirror.

Items		Specifications	
		Inside	Outside
Shape and dimensions	Mirror size width (torsion bar side) × length (other side)	3 × 4 mm	
	Dimensions	50 × 35 × 9 mm	
	Mass	43 g	
Basic performance	Resonance frequency	1500 Hz ± 5%	440 Hz ± 5%
	Amplitude at resonance frequency (optical angle)	Less than ±30°	Less than ±30°
	Frequency–amplitude characteristics (sine wave driving)	See Figure 10.9	
	Current–amplitude characteristics with amplitude at resonance frequency (sine wave driving)	See Figure 10.10	
Optical characteristics	Material of mirror	Au	
	Mirror reflectance	More than 85% ($\lambda = 670$ nm)	
Electrical properties	Value of resistance	230 Ω ± 10%	410 Ω ± 10%
Lifetime Number of scans		More than 5×10^9 times	

Since the Eco Scan is driven by an electromagnetic method, the current is almost proportional to the amplitude. Furthermore, as the Eco Scan has a high Q-value (Q-factor), a large amplitude can be obtained around the resonance frequency, but the amplitude falls to as low as 0.1° at a non-resonance driving frequency.

If the moment of inertia of the mirror part is J, the spring constant of the supporting part is k and the resonance frequency is f, the resonance frequency can be obtained with the following equation:

$$f = \frac{1}{2\pi}\sqrt{\frac{k}{J}} \tag{1}$$

The relation between the drive power M and the amplitude θ is

$$M = k\theta \tag{2}$$

From both Eqs. (1) and (2), the amplitude θ can be obtained from

$$\theta = \frac{1}{4\pi^2}\frac{M}{f^2 J} \tag{3}$$

This means that when the drive power M is constant and if the resonance frequency becomes high, the amplitude will become smaller by the second power of the resonance frequency, and if the size of mirror part becomes larger, the amplitude will become smaller by the first power of the moment of inertia of the mirror part.

As explained above, to drive the Eco Scan effectively it is necessary to design the operating frequency to be as close to the resonance frequency as possible, considering the mirror size, the resonance frequency and the amplitude.

Frequency responses and rotation amplitudes of the two types of device are shown in Figures 10.7–10.10.

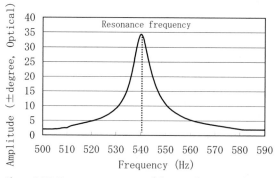

Figure 10.7 Frequency response of the one-dimensional type of mirror device at resonant frequency. The amplitude is scaled by the optical reflection angle. The device is driven by a sinusoidal wave.

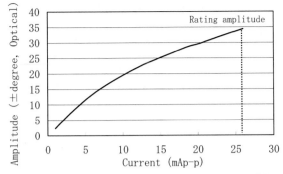

Figure 10.8 Rotation amplitude at resonance frequency of the one-dimensional type of mirror device against the driving current. The device is driven by a sinusoidal wave.

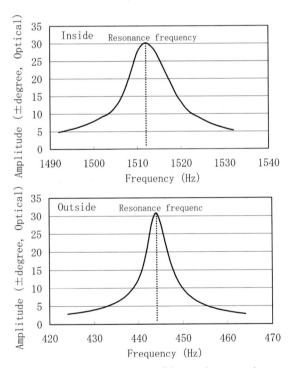

Figure 10.9 Frequency response of the two-dimensional type of mirror device at resonant frequency. The amplitude is scaled by the optical reflection angle. The device is driven by a sinusoidal wave.

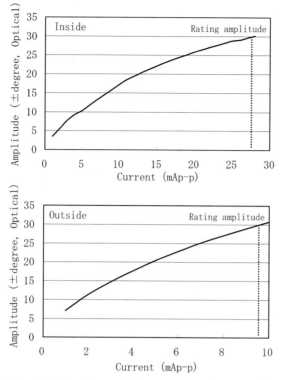

Figure 10.10 Rotation amplitude at resonance frequency of the two-dimensional type of mirror device against the driving current. The device is driven by a sinusoidal wave.

For reference, the specifications of the developed Eco Scan devices of the one- and two-dimensional types are given in Tables 10.4 and 10.5, respectively. Pictures of the mirror and coil arrangements are shown in Figures 10.11–10.13.

10.6
Material for Optical Mirrors

The mirror of the Eco Scan is formed on the surface of a silicon wafer by coating, so the most suitable type of coating can be selected matching the laser wavelength to be used. In the case of a wavelength of >650 nm such as an infrared laser beam or red laser beam, a gold mirror is the best, and with a shorter wavelength such as a green or blue laser beam, an aluminum or silver mirror is the most suitable.

Table 10.4 Specification of developed one-dimensional type mirror devices.

Mirror size (mm)	Resonance frequency (kHz)	Amplitude (optical angle) (°)	Notes
5 × 6	0.5	±34	Double-sided mirror See Figure 10.1
4 × 4	1	±18	See Figure 10.11
4 × 4	2	±20	
4 × 4	4	±12	
4 × 4	8	±8	
Φ-1	D.C. driving	±1 (stoppable at any position)	Dimensions: Φ 5.5 × 10 mm See Figure 10.12

Table 10.5 Specification of developed two-dimensional type mirror devices.

Mirror size (mm)	Inside		Outside		Notes
	Resonance frequency (kHz)	Amplitude (optical angle) (°)	Resonance frequency (Hz)	Amplitude (optical angle) (°)	
3 × 4	1.45	±30	440	±30	See Figure 10.2
Φ-3	6.3	±15	D.C. drive	±3.5	See Figure 10.13

Mirror with the coil on its back

Figure 10.11 One-dimensional type of resonant optical mirror device. The mirror is 4 mm square.

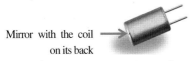

Mirror with the coil on its back

Figure 10.12 One-dimensional type of resonant optical mirror device. The mirror is 1 mm in diameter.

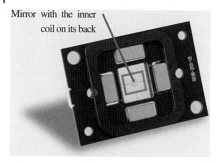

Mirror with the inner coil on its back

Figure 10.13 Two-dimensional type of resonant optical mirror device. The mirror is 3 mm in diameter.

10.7
Reliability

10.7.1
Static Strength [5]

The torsion bars that support the mirror part are twisted largely by Lorentz force in operation. Consequently, the torsion bar made of single-crystal silicon may fracture due to its brittleness. Since single-crystal silicon does not have any internal defects, the strength depends on the short cracks on the surface. The short cracks arise mainly due to the etching process and the effect is referred to as "etching damage".

10.7.1.1 Etching Damage

As explained in Section 10.4, in the manufacturing process an SOI wafer is used for the Eco Scan and it is processed into the desired shape by inductively coupled plasma reactive ion etching (ICP-RIE). ICP-RIE enables a deep groove to be formed in the wafer by repeatedly forming a protective film using octafluorocyclobutane (C_4F_8) and etching with sulfur hexafluoride (SF_6). Also by using an SOI wafer, the etching stops just at the silicon dioxide (SiO_2) interface and it is then easily possible to form the desired shape (Figure 10.14). However, through the ICP-RIE process, usually fairly extensive etching damage (surface roughness) remains on the lateral side (etching surface). Three kinds of typical etching damage caused by ICP-RIE are mask, side-wall and notching damage, as summarized below. Figure 10.14b–d show the results of lateral surface etching damage as observed by scanning electron microscopy (SEM).

    1. **Mask damage.** The damage marked A that occurs on the lateral side at the beginning of etching on the upper part (Figure 10.14b) and the part marked A in Figure 10.14a is caused by erosion of the resist mask. The edge of the resist mask coated on the upper surface of the wafer is gradually

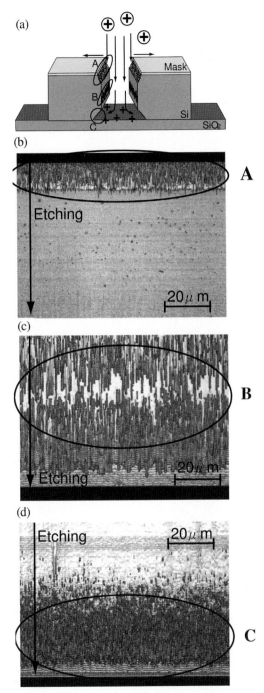

Figure 10.14 (a) Schematic of the process of etching damage nucleation. (b)–(d) SEM images of etching surface (side-wall of trench). The etching surface is subjected to various kinds of heavy damage. The types of damage A, B and C shown circled are referred as mask damage, side-wall damage and notching damage, respectively.

etched during ICP-RIE and the exposed upper wafer surface is damaged.

2. **Side-wall damage.** In the area marked B on the upper part in Figure 10.14a, at the end of etching, roughness of the lateral side (side-wall damage) is observed (Figure 10.14c), where horizontal and vertical wave patterns can be seen. The former are caused by repetition at short intervals of both the etching and passivation processes to form a protective film on lateral side and one cycle of the wave pattern corresponds to one gas switching process. The latter is considered to be the roughness caused by uneven etching of the lateral side protective film and by the partly delayed etching due to the uneven etching. If this happens, the record is left in the direction of etching and columnar patterns are formed.

3. **Notching damage.** The damage observed near the end of lateral side etching (Figure 10.14d) (at the SiO_2 interface) is called notching (the area marked C in Figure 10.14a). The mechanism of notching is as follows. The electrons in the plasma only reach as far as the wafer surface because of the ion sheath formed in the plasma on the SOI wafer. Therefore, the electrons hardly reach the bottom of the narrow pattern to be made. Consequently, the bottom of the pattern to be made is charged positively by positive ions, and the ions, the etching species, react and bend and enter the lateral side of the pattern and cause damage. This is called notching or footing and the damage occurs at the end of the process in the direction of etching. In general, this notching damage is the largest type of damage in this process. In addition, a special kind of wave pattern is observed at the lowermost part. This is considered to be caused by a notching-free process.

Among the above types of damage, especially notching could lead to problems with actual equipment, such as degradation of performance or deterioration of stiffness, and notching is considered to have a large influence on the intensity. Accordingly, it is an important issue to evaluate the intensity of the notching part and to increase the intensity there.

10.7.1.2 Intensity Evaluation of Notching Part

The relationship between the average roughness (R_a) of the notching parts of multiple samples from different etching processes and the fracture strength obtained by the bending test is shown in Figure 10.15. Additionally, R_a was measured with a laser microscope.

As can be seen in Figure 10.15, a linear relationship is observed between the notching roughness and the intensity, and it was found that the intensity could be improved by ~230 MPa or more if the roughness was improved by 0.1 μm.

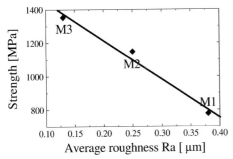

Figure 10.15 Relationship between scale parameter (fracture strength) and roughness of notching region.

Also, after having improved the notching roughness, other etching damage such as mask damage becomes dominant, so care must be taken when interpreting the intensity evaluation.

10.7.2
Fatigue Life [6]

The Eco Scan is formed on single-crystal silicon as a single piece, so metal fatigue will never arise, and it has no bearings, which polygon mirrors and galvanometer scanners usually have, so there will be no abrasion. Therefore, the lifetime of the Eco Scan is very long. We have been driving a one-dimensional type (Figure 10.1) for 24 h per day continuously for two full years and it is still working without any problems.

10.7.2.1 Designing Fatigue Life
To estimate the lifetime of the Eco Scan under a normal environment, the following equation is used for the evaluation of the fatigue life, L_f:

$$L_f = AP^{-B} \exp\left(\frac{C}{T} - D\varphi\right) \tag{4}$$

where A, B, C and D are the parameters to be used for the experiment, P (%) is the relative humidity, T (K) is the temperature and φ (°) is the optical angle. Concerning the relative humidity P, it is assumed that the fatigue life is proportional to P to the power of temperature, drawing upon examples of the empirical equations used for many kinds of electronic devices:

$$L_f \propto P^{-B} \tag{5}$$

We assumed that the temperature T could be adjusted through the general process of thermal activation and used an Arrhenius model according to the equation

$$L_f \propto \exp\left(\frac{C}{T}\right) \tag{6}$$

Concerning the stress, Eq. (7) is used, which assumes that the logarithm of fatigue life is proportional to the stress. Additionally, the optical angle φ (°) is considered to be proportional to the stress.

$$L_f \propto \exp(-D\varphi) \tag{7}$$

10.7.2.2 Fatigue Life Test Results

The fatigue experiment was performed using the Eco Scan shown in Figure 10.1 with a resonance of 500 Hz. The experiment was carried out under the conditions of optical angles of ±34, 40, 50, 55, 60, 70 and 80°, temperatures of 20 and 85 °C and relative humidifies of 25 and 80%.

From the data obtained, the values of A, B, C and D in Eq. (4) were obtained. Figure 10.16 shows the fatigue experimental data obtained from an experiment performed under an accelerated life environment with a temperature of 85 °C and a relative humidity of 80% and the 99% prediction interval. All the experimental data are within the 99% prediction interval.

Figure 10.17 shows the lowest values of the 99% prediction interval related to the optical angle and the lifetime under the reference atmosphere of Japan (average highest temperature 19.7 °C and average relative humidity 63%) as follows. Targeting a lifetime of 10 years (5 × 10^8 times), it was found necessary to have an amplitude with an optical angle of about ±35°, and in the case of 100 years (5 × 10^9 times) an amplitude with an optical angle of about ±25° would be required.

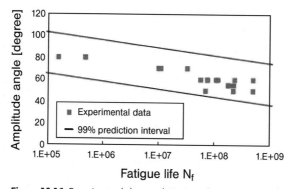

Figure 10.16 Experimental data and 99% prediction interval of the fatigue life of the MEMS micromirror (temperature 80 °C, relative humidity 85%).

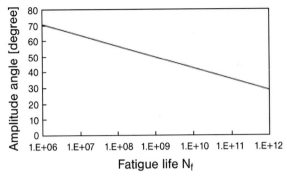

Figure 10.17 Lowest 99% prediction interval of the fatigue life of the MEMS micromirror (temperature 20 °C, relative humidity 64%).

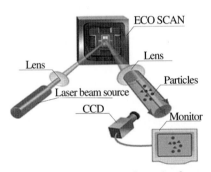

Figure 10.18 Visualization of particles floating in the air.

10.8
Applications

Some examples of the application of the Eco Scan are outlined below. In addition to these examples, the demands on light scanning such as with optical switches or optical attenuators for the information technology (IT) industries are great in many different fields.

10.8.1
Visualization of Particles Floating in the Air (Figure 10.18)

By scanning laser light with the Eco Scan at high speed and covering a wide range, particles floating in the air, which are of micron size, can be visualized by capturing the scattered light reflected from the particles.

10.8.2
Measurement of Shapes of Objects (Figure 10.19)

As shown in Figure 10.19, the object is exposed to a laser beam scanned by the Eco Scan. By detecting the light reflected from the object while it is rotating, the shape of the object can be measured without touching it, i.e. non-contact measurement.

10.8.3
Blocking-type Area Sensor (Figure 10.20)

By arranging two Eco Scans and two retro reflectors as shown by Figure 10.20, they can be used as blocking-type area sensors. By installing this type of area sensor on the platform of a railway station, it can detect human bodies or objects that protrude from the platform.

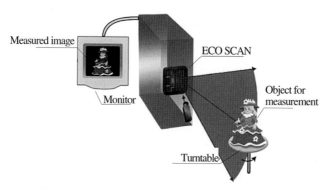

Figure 10.19 Measurement of shapes of objects.

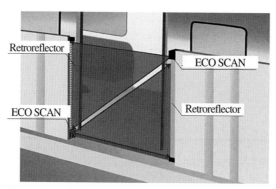

Figure 10.20 Blocking-type area sensor.

10.8.4
Vehicle-mounted Laser Radar (Figure 10.21)

With a two-dimensional Eco Scan mounted on the front of a car, any cars or obstacles ahead can be detected by light scanning with the Eco Scan.

10.8.5
Laser Printer (Figure 10.22)

By scanning the laser beam entering the Eco Scan over a photosensitive drum and by synchronizing the laser beam with the Eco Scan, characters or pictures can be displayed on the drum. Printing can be performed by transferring such characters and pictures on to paper.

10.8.6
Barcode Reader (Figure 10.23)

The barcode is exposed to a laser beam scanned by the Eco Scan and the barcode can be read out.

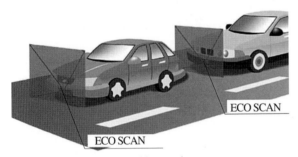

Figure 10.21 Vehicle-mounted laser radar.

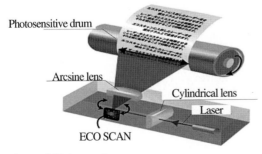

Figure 10.22 Laser printer.

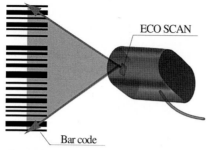

Figure 10.23 Barcode reader.

Figure 10.24 Ranging sensor.

10.8.7
Ranging Sensor (Figure 10.24)

By measuring the time for the laser beam to reach, to be reflected by and to return from an object, the distance between the object and the Eco Scan can be measured. By combining this distance information with the two-dimensional scanning information obtained by the Eco Scan, the three-dimensional shape of the object can be detected.

10.8.8
Laser Display (Figures 10.25–10.27)

It is possible to write letters or draw pictures on a screen by synchronizing a laser beam with the two-dimensional Eco Scan (Figure 10.25). With a laser beam and the Eco Scan, the optical system can be downsized, so it becomes possible to apply the system to mobile phone laser displays (Figure 10.26) or head-up displays (Figure 10.27) for automobiles.

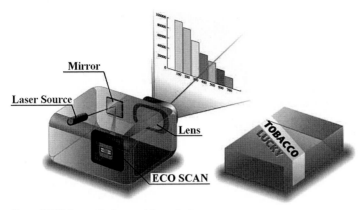

Figure 10.25 Laser display: mobile projector.

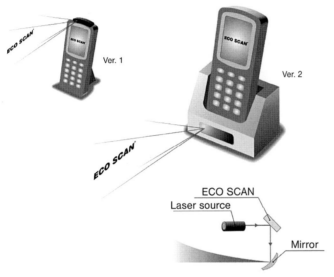

Figure 10.26 Laser display: mobile projector with cell phone.

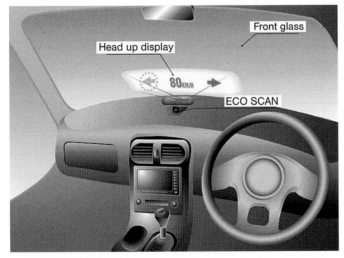

Figure 10.27 Laser display: head-up display.

References

1 Texas Instruments Incorporated, [http://www.tijz.co.jp/jrd/dlp/docs/index.htm].

2 N. Asada, H. Matsuki, K. Minami, M. Esashi, Silicon micromachined two-dimensional galvano optical scanner, *IEEE Trans. Magn.* **30** (1994) 4647–4649.

3 N. Asada, M. Takeuchi, V. Vaganov, N. Belov, Silicon micro-optical scanner, in *Digest of Technical Papers, Transducers '99, Sendai*, 778–781, 1999.

4 The Nippon Signal Co., Ltd., [http://www.signal.co.jp/vbc/mems/].

5 S. Izumi, M. Yamaguchi, K. Sasao, S. Sakai, Y. Ueda, A. Suzuki, Strength evaluation for notching damage of MEMS micromirror by dual-direction bending test, *Transactions of the Japan Society of Mechanical Engineers, Sevies A*, Vol. **72-717** (2006) pp. 720–727.

6 S. Izumi, M. Kadowaki, S. Sakai, Y. Ueda, A. Suzuki, Proposal of a reliability-based design method for static and fatigue strength of MEMS micromirror, *J. Soc. Mater. Sci. Jpn.* **55** (2006) 290–294.

Index

Reliability of MEMS.
Edited by O. Tabata, T. Tsuchiya
Copyright © 2013 WILEY-VCH Verlag GmbH & Co. KGaA, Weinheim
ISBN: 978-3-527-33501-5